C1540

The Ladder of His Life

Biography of

Air Chief Marshal Idris Hasan Latif, PVSM

The Ladder of His Life

Biography of
Air Chief Marshal Idris Hasan Latif, PVSM

Bilkees I. Latif

KONWLEDGE WORLD
KW Publishers Pvt Ltd
New Delhi

in association with

Centre for Air Power Studies
New Delhi

2011 FPBA BEST PUBLISHERS AWARD (ENGLISH)

The Centre for Air Power Studies is an independent, non-profit, academic research institution established in 2002 under a registered Trust to undertake and promote policy-related research, study and discussion on the trends and developments in defence and military issues, especially air power and the aerospace arena, for civil and military purposes. Its publications seek to expand and deepen the understanding of defence, military power, air power and aerospace issues without necessarily reflecting the views of any institution
or individuals except those of the authors.

Jasjit Singh
Director General
Centre for Air Power Studies
P-284, Arjan Path
Subroto Park
New Delhi 110010

Tel.: +91.11.25699131
E-mail: diroffice@aerospaceindia.org

Published in India by
Kalpana Shukla
KW Publishers Pvt Ltd
4676/21, First Floor, Ansari Road, Daryaganj, New Delhi 110002, India
Tel.: +91.11.23263498 / 43528107 Email: knowledgeworld@vsnl.net
Website: www.kwpub.com

ISBN 978-93-81904-85-5

Printed and bound in India

Dedication

This book commemorates
Idris' 90th birthday on
9th June 2013

I dedicate this book to our
three lovable children:
Mariam, Asad and Asgar and
their wonderful partners
as well as their lovely children
who have all brought so much
happiness into our lives.

Contents

Preface

June 2013

Just this month, I celebrated my 90th birthday. At my age, one tends to frequently reflect upon the life that has passed. Happily, most of my days are spent with a smile, having been blessed with a life so full of good fortune and happiness. For this, there are so many to whom I am deeply indebted. I have been blessed with Bilkees, my beautiful and selfless and loving wife, with a heart of gold, who has stood by my side steadfastly for the last sixty-two years, and who has, like no other I know, laboured tirelessly for the last year and more to complete this work. To my three children who have each contributed to this work with their thoughts and ideas, and, with their ever caring and loving ways, constantly bring so much sunshine into our lives.

I was blessed to have been born into a wonderful family and provided with unconditional love and the opportunity of education. To develop a passion for poetry that brings me comfort even in moments of distress. To serve for forty years in the Indian Air Force (IAF) with a group of men who became my friends, my comrades, in whose hands my life rested and who entrusted theirs to me. They became our family and we theirs. I am grateful to have had the opportunity to serve as Chief of the Air Staff and pleased that much was achieved on my watch. The credit goes in no small measure to all those who I was privileged to serve with. Later in my life, both at the Governor's office in Mumbai and also at our Embassy in Paris, once again, so much is owed to all those who were always there to offer their selfless support. Bilkees has written this book to record all that we have been blessed with but I must add how much the participation of so many of my superiors, my colleagues and my subordinates, both civilian and those in uniform, many truly outstanding, made all this possible.

The world today is perhaps a little more complex than mine was when I was my grandson's age. As a fighter pilot in World War II, we woke up well before dawn, flew by the seat of our pants, raced motorcycles and some of us came through unscarred. In so many ways, our lives were very simple, with few expectations. Demands on the individual today are perhaps greater, and our society appears more concerned with "me" and less with "us". Yet, as can be testified by the experiences of my life, I still see the glass half full and not half empty, and I have no doubt that the basic culture, the fine traditions and the selfless outlook of our great people, both tolerant and intelligent, will prove more than equal to the challenges that lie ahead for all of us.

Jai Hind

Idris Latif

CHAPTER

01 Who is Idris Latif?

Who is Idris Latif? If he is to be formally addressed on any occasion, should it be as, 'His Excellency' as he is sometimes addressed, having been the Governor of a state? Or as Ambassador, since he was Ambassador Plenipotentiary with Minister of State rank during his three years in Paris? Or as Air Chief Marshal, the rank he finally held in his beloved Air Force? This last, I know, is by far his preference — but that also only if a formal occasion so demands.

Or, Idris Latif, who has shouldered varied responsibilities as a dedicated patriot; a Fighter Pilot, a devoted former Chief of the Indian Air Force with its outstanding personnel who were, and are, totally committed to our nation — these personnel at Air Force bases, units and headquarters located all over India — loyal and devoted, never grudging the losses of their own young lives that are the inevitable consequences of duty in the armed forces, particularly, in the Air Force, even in times of peace?

Or, Idris Latif, selected in the early 1970s, as the very first newly created Assistant Chief of Staff (Plans), and made directly responsible for planning the future shape and size of key elements of the Service? The DPSA (Deep Penetration Strike Aircraft) programme, based on the Jaguar (today a key element of our strike force), was initiated by him, and finally entered into service when he was the Air Force Chief.

Or, Idris Latif, who did so much towards the morale of Servicemen of all ranks in the Air Force — both officers and men — which had been affected due to poor promotion prospects? And who, from his very first day in office as the Chief, worked

tirelessly to improve upon the human resource policies for the benefit of all personnel, and with it, importantly, improving thereby the operational capability of the Air Force?

Or Idris Latif the pilot, who fully qualified himself and flew no less than 22 different types of aircraft, including fighters, bombers and even heavy transport aircraft; and who, even in the final years of his service as the Chief, found time to fully qualify himself on the Sukhoi-7, the MiG-21 and the Jaguar; plus make time for familiarisation sorties on the Mirage 2000 (during an official visit to France), and, finally, the MiG-23 and even the MiG-25 just months after the aircraft first arrived and were being assembled, before his final retirement?

Or, Idris Latif, who, in his forty years of service, spent almost two-thirds in Operational Units and formations and Staff appointments at Air Headquarters which had a direct bearing on the operational readiness of our Air Force — Idris Latif who, in his final three years as the Air Chief, achieved perhaps the most significant modernisation of key elements of our Air Force?

Or, Idris Latif, who, perhaps because of his forty years of man-management in the Air Force, coupled with his integrity, loyalty and concern for the nation and its people, was so effective in his next appointment as Governor of Maharashtra?

Or, finally, the Idris Latif whose insistence on excellence in every sphere, his friendliness and charm, quick reactions and wit, elegant taste and natural warmth, made him, not only an effective Ambassador of India in France (his final government appointment), but with it, an able and effective organiser of the Festival of India?

I write about Idris Latif who has been ALL these and still is much, much more to his family and to his friends. I have tried over the last few years to get him to speak about Service life, about which we would all love to know more. He would, however, never speak on any Service matter, particularly if he felt that, at a particular point in time, security would in any way be compromised. Our children, especially the two boys, have also been very deeply interested in life in the Air Force; in fact, our younger son had got selected for the Air Force. It was quite often when the children and Idris and I were relaxing together that they would ask Idris about various episodes and which, if security considerations permitted (about which we often felt he was excessively concerned!), he related to our fascinated three children and to me. I would often jot

down quick notes at the time and have used them; and, finally also, I have to say, on several occasions, I have spoken to others for this book. Idris has insisted that I should add material about myself, as, according to him, once we were married, "She was also a member of our Air Force, and over the years, on many, many occasions, proved to be a pillar of strength." I have, therefore, with some reluctance, added accounts of some of my own far too numerous involvements.

Born on 9th June 1923, Idris joined the Air Force on 1st September 1941 and was commissioned on 26th January 1942 as Pilot Officer No: 1804 GD(P) of the Indian Air Force or, more accurately, the Indian Air Force Voluntary Reserve, as it was initially called. Following years of devoted and competent service, he was appointed Chief of the Air Staff on 1st September 1978 and retired on 31stAugust 1981, after exactly 40 years in his greatly loved Air Force. He was then appointed the Governor of the State of Maharashtra in March 1982 and, concurrently, for some time, the Lieutenant Governor (as it was then) of Goa. After three exciting and challenging years in Maharashtra, he was appointed Ambassador Plenipotentiary to the Republic of France with the rank of Minister of State for three years from April 1985.

Idris also had to decline the medal being awarded to him by the United States, after his tenure as Air Attaché at our Embassy in Washington D.C. He was informed that acceptance of such awards was against our government's policy, at least at that time. Noteworthy also, I feel, is that Idris, in his long career of almost 50 years, never asked, or tried in any way for any appointment or position. Nor did he attempt to get out of any posting that did not attract him, for instance, a prolonged absence from active flying assignments. Offered other special appointments after returning to Hyderabad after his 50 years in government service — even another stint as Governor — he chose not to take on further official responsibilities, primarily — in fact, solely, I think — so that he could spend time with his widowed, aged and bedridden 90-year-old mother in Hyderabad. We have, therefore, ever since, continued to live in Hyderabad and do so with no regrets.

Among Idris' ancestors was Sir Akbar Hydari, the late Prime Minister of the Dominions of Hyderabad. His great uncle, Justice Badruddin Tyabji was the third elected President of the 3rd session of the Indian National Congress (Madras, 1887). Abbas Tyabji, nephew of Badruddin Tyabji, and of Idris' great grandmother, was Chief

Justice of the Baroda High Court, who gave it up to join Mahatma Gandhi. During the Freedom Movement in India, on 4th May 1930, after the Salt March to Dandi, Gandhiji was arrested. Gandhiji had nominated Abbas Tyabji in charge of the next phase of the Salt Satyagraha, and it was Abbas Tyabji who then led the raid on the Dharasana Salt Works in Gujarat. Like others of his family, Idris, too, was always willing to give his life in the service of our nation.

Idris' suggestions to the Chief of the Air Staff, Air Chief Marshal Lal (kept top secret at the time) for rendering enemy airfields temporarily unusable, were implemented and used successfully in the 1971 War. Idris was at Air Headquarters in the newly created post of Assistant Chief of the Air Staff (Plans) in the rank of Air Vice Marshal. It was in this capacity that he also carried out the demanding tasks of making first-hand assessments of frontline combat squadrons and the modernisation plans of the Air Force. For his role, Idris received the Param Vishist Seva Medal in 1971.

Idris was in the Eastern Sector at Shillong during the 1971 conflict when the final phase of the operation leading to the surrender in (then) East Pakistan took place. He had been especially directed by the Chief of the Air Force himself, to be available at all times. He was also to, at very short notice, if the situation demanded, personally organise and monitor, in consultation with the Air Force formation directly concerned, any task that he (the Chief) may entrust to him. On a very memorable occasion, when at Shillong (the Headquarters of Eastern Air Command), on Air Chief Marshal Lal's personal telephonic directions at 8 o'clock one morning, Idris became directly involved in organising the exceptionally accurately timed and superbly executed rocket attack by our fighters, on the Circuit House in Dhaka. Idris had apparently been told by the Air Force Chief that an urgent meeting of the Governor's Council had been called there by the East Bengal Governor for a major policy decision, and the timing was of critical importance. Two independent sources confirmed later that the rockets struck the building with exceptional accuracy. Just minutes later, as a direct consequence, the Pakistanis decided that their forces should unconditionally surrender (more on this in Chapter 15). Our Army had fought brilliantly and it was rightly appreciated and given full credit. But it was widely acknowledged that the contribution of the Air Force, under the inspiring leadership of Air Chief Marshal Lal, was exemplary

in every way. Indeed, it was of critical importance to the quick and successful outcome of the war.

On his return to Air Headquarters at the conclusion of the war, Idris, as Assistant Chief of the Air Staff (Plans), authored a detailed report on the lessons learnt for planning the future of the Air Force; in particular, the modernisation of its strike element by the acquisition of combat aircraft, including the Jaguar.

Finally, perhaps just a year before his retirement, Idris initiated the Light Combat Aircraft (better known as the LCA). Recently, he explained to me its critical importance: it was high time, he had strongly felt, to once again try and reverse our almost total dependence on imports for our Air Force's strike force, particularly considering the government's enormous investments of national resources in Hindustan Aeronautical Limited (HAL). Secondly, it was necessary to ensure a timely replacement for our ageing MiG-21 fleet. Also, keeping in mind the continuing paucity of funds and HAL's limitations, particularly in design and development, we should not aim for a sophisticated aircraft and, consequently, a much more expensive one. Calling it the LCA would, thus, help underline the critical importance of the LCA being a Low Cost Aircraft also. Idris said that his senior-most Air Force Commanders and Staff, Chairman HAL and the Defence Ministry (including the Scientific Adviser to the Defence Minister), all agreed, and the LCA project, happily, became a unanimous decision.

Right up to the end of his career in the Air Force, Idris' love for flying remained. He always, like most other pilots of the Air Force, wanted to fly and fly! One of his last acts before retirement was to fly the tri-sonic MiG-25, which had been assembled on high priority by Air Force personnel from the semi-knocked down condition in which it had been dispatched from, what was then, the Soviet Union.

The children and I never got involved in, nor discussed, any of his assignments. Once I was in Mumbai in 1971 at a party when one of the guests asked me something about one of our aircraft. I was completely ignorant about it and did not even recognise the name of the aircraft he mentioned. He laughed when I said that my husband never discussed these things with me, and added that there were crates full of spares of these aircraft with their name on them, all lying at the docks for everyone to see, so why was there such secrecy? However, that was how it was. Service matters were not

discussed with the families, which was and is a good policy. I found our stay in various places certainly challenging, but each also soon very enjoyable. In fact, even following our marriage, the transfers were so frequent that finally we chalked up a total of 28 homes that we moved to.

After 57 years of married life together, I decided to write this book in consultation with our three children, Mariam, Asad and Asgar, in spite, I have to add, of Idris' great reluctance to having a book written about him. He has always shunned publicity for himself — except when it is linked to praise for our Air Force! We, the family, consisting of our children and grandchildren, however, felt that his life should be recorded — not only as Chief of the Indian Air Force, Governor of Maharashtra and Ambassador of India to France with Ministerial status, but as a man of true values.

Idris is warm, loving, caring and quick-witted, with an easy sense of humour. He insists strictly on discipline, thoroughness and perfection in his own actions and in the actions of those around him. He is a human being with a quick brilliance, absolute integrity, patriotism and dedication to the nation, and a stickler for punctuality and duty, as well as correct Service protocol. He is an instinctive gentleman and never uses foul language or swear words, even when he is in a temper. Idris is a man of many interests and many talents. For Idris, his life in the Air Force, which he served for forty years, was his true vocation. Till today, this Service remains his deepest involvement. Yet he fitted in perfectly as Governor of Maharashtra, and again as Ambassador of India to France.

Idris has had infinite faith in God, and has no doubt that He would never, ever forsake him. However, I do know that his last few years in Service, before he was appointed Chief, were his most difficult ones. Still, I will not write about all that he had to face, particularly as, with Idris, his principles, and foremost amongst them is loyalty to the Service (which includes subordinates and superiors alike), are unshakable. Speaking ill of a superior, particularly in public, we have both always agreed, is, in fact, being disloyal to the Service. Idris is a perfectionist and whatever he does is invariably completed with a real touch of excellence.

It was a great honour that Idris' name came up for consideration for President of our country. The first time was while he was Ambassador Plenipotentiary to France. We got to know of it when we received newspaper cuttings from Maharashtra. One of the cuttings is shown here.

THE NEXT PRESIDENT?

WHO will be the next President of India? The question has begun to reverberate in the Capital's corridors of power. Names are being mentioned in various circles. Three stand out: Our Ambassador in Paris Idris Latif, Union Minister B. Shankaranand and Trade Fair boss Mohammed Yunus, in that order.

CURRENT Exclusive

The near consensus in the ruling circles is that the eighth President of the Republic would, and should, for obvious reasons, be either a Muslim or a Harijan of eminence. And the somewhat personal factor in most calculations involving the Gandhis is that the

Giani Zail Singh is almost certainly out of the race as the nominee of the ruling party. The differences between him and Rajiv Gandhi on vital issues including the notorious Postal Bill have widened the gulf between them. Whether the

social grace, pushing him up as an elder statesman. While this should make him ideally suited for elevation, his advancing age and his caste stand in the way. He is a Brahmin.

From the south Chief Minister M.G. Ramachand-

Muslim for President' idea are projecting Union Minister Mohsina Kidwai, coming as she does from the Hindi heartland and the famous Kidwai clan. And the country can have the first woman head of state, which was not possible during Indira

been tarnished by a seri scandals and the P Minister, despite fondness for him, may take the risk of considering his name.

Gujarat, Rajasthan, Ke Kashmir, Orissa, Hima and Haryana continu draw a blank. The probable candidat Speaker Balram Jakhar circles close to him r that he is "very keen" to Rashtrapati Bhavan, he is from Punjab, to w the present incumbent belongs.

It leaves among top n Arjun Singh and Siddh Shankar Ray, both h cultivated personali Despite his meteoric rise sudden fall, Arjun Singh be only too glad to oc the highest office. But the Ganga will be on Ray, like Brahmana Reddy, carries the cro the Shah Commission. Rajiv sticks to the poli forgive and forget Ray very well be the choice. boost the Bengali ego will certainly affect Ma in the east.

Karnataka's Shank nand does not measure the stature of Ray. His qualification is that he is the deep South, is a Ha

A press cutting received while in Paris, late 1980s

While Idris was serving in France, he was asked in very great confidence by Prime Minister Rajiv Gandhi if he would agree to be the Vice President of India. Idris said that he was deeply, very deeply, touched and highly honoured to be considered for such an office. However, he sincerely felt temperamentally unsuited for the position, particularly as his insistence on a very high standard of discipline, which has become so much a part of him, may well prove to be, politically, in the Rajya Sabha, a cause of serious embarrassment for the government.

We are both now happily settled in Hyderabad, Idris involved in his hobbies and family, helping others, when he can. He has been offered several assignments but prefers not to take on more responsibilities. Our time together has been very

precious. We share many views and interests, but have very different ones too, so that our years together have never been dull, in fact, always stimulating. Our years together here in Hyderabad have been precious, and our greatest boon has been our wonderful children, their lovable spouses and most beautiful five grandchildren, with their values, integrity, caring and qualities of head and heart.

The following will give an idea of Idris' outlook and thoughts through his Address to the Air Force Stations on taking over as the Chief of the Air Staff.

Over the years, our Air Force has grown enormously and our Stations, Squadrons, Units and Depots are, today, spread right across the length and breadth of India. My visit to this station, therefore, has necessarily to be somewhat brief. And it will not, unfortunately, be possible for me to meet too many of you individually, but I am happy at the opportunity to speak to you at least collectively, so that I may share with you some thoughts that are uppermost in my mind. This visit will also enable me to see for myself the advancements of your hard work and, at the same time, give me an opportunity to familiarise myself with the problems and difficulties that you face so that we may, jointly, examine them and seek solutions and remedies.

First and foremost, I think it is important that we constantly bear in mind and remind ourselves and appreciate, that the equipment, the facilities that have been procured for our Air Force and placed at our disposal, in our charge, are truly of staggering dimensions, costing hundreds and hundreds of crores of rupees — procured for us at a cost that our country can ill afford. It is our responsibility, yours and mine, our joint responsibility to ensure that we make the best possible use of it. It is also necessary that we remind ourselves that the strength and capability of our Air Force, of any force or organisation for that matter, depends, not only on what we have, but on what we do with what we have.

Our inventory of weapon systems is, by and large, balanced, sophisticated and up to date. Our training institutions are, by now, well established and

cover the complete spectrum of the requirement of any Air Force; undoubtedly, no developing country can boast of a supporting infrastructure, including facilities for maintenance and overhaul, such as ours. Let us recognise that the government has discharged its responsibility; the rest is up to us.

And we in India, we in our Air Force, have another tremendous asset: the technical and specialist skills of our personnel, be they officers or men, Serviceman or civilians. In skills and in experience, in ingenuity and in enterprise, we are second to none. If the will is there, the determination is there, the resolve is there, there is nothing that we cannot accomplish.

If the equipment is there, the facilities are there, the technical and specialist skills are there, what else is required? It is hard work, dedicated and determined hard work. It is, gentleman, our capacity to work in times of peace, and the quality of that work, that will determine, without doubt, the measure of success we achieve in war — should hostilities be thrust upon us once again. Never forget that the more we sweat in peace, the less we bleed in war.

The long history of war has proved, if nothing else, the supreme importance of the human factor, the human element, in the total weapon system. What ultimately decides the outcome of any battle, of any war, is the quality of the man — his training and his morale. Happily, our Air Force, in both strength and sophistication, is more than adequate to meet any challenge. If our thinking and our training remain positive and progressive, and operationally oriented, we will remain invincible. Let us resolve to leave no stone unturned to make it so.

Whilst I call upon you to make this effort, I would like to take this opportunity to assure you that the problems that you have, the limitations under which this and other stations have to function, or the difficulties that many of our personnel face, will not be forgotten. I know, for instance, that shortage of married accommodation is a major problem; the schooling of our children is another. Your Station Commanders, your AOC-in-C and I and, indeed, the government, all are deeply conscious of this. I can assure you that all of us, at every level, will continue to make sustained efforts

to quicken the pace of new construction and quicken, also, the pace of improvement and expansion of existing facilities.

As the same time, when looking at our problems, it is only right that we should also look at the problems that millions of our countrymen face. We then see our own difficulties, our own problems, in proper perspective. Look back also on what we were, and what we are today. How we lived and worked when Partition came, and that was not so long ago. Ambala is perhaps the only permanent station that came to us; all others went to Pakistan. Today, in whichever direction you turn — from Srinagar in the north, to Bangalore and Madras in the south and from Chabua in the east, to Jamnagar in the west — we have innumerable permanent stations, well built, well equipped with all kinds of welfare facilities — facilities that we did not even dream of when we joined the Service.

But remember, also, that camps and houses are not built by brick and mortar alone; each one of us has a contribution to make. As we, today, enjoy the fruits of the initiative, the enterprise and the hard work of those that have gone before us, let us resolve to play our own part in full. And if each one of us does that, thinking less of oneself and more of the future, genuinely and sincerely, our Air Force, without a shadow of doubt, will continue to grow from strength to strength and remain the pride of our people and our nation.

Thereafter, for three years, Idris was the Chief of his most precious and beloved Air Force. I have reproduced below a special treasure from the Prime Minister of India, of the time when Idris laid down the mantle of Chief of our Air Force. It is the only time, to our knowledge, that such a release was issued.

The Prime Minister's press release re: Air Chief Marshal I.H. Latif

NEW DELHI, September 1 (PTI): The Prime Minister has highly, appreciated the manner in which Air Chief Marshal I.H. Latif, who retired yesterday, headed the Air Force, raising the morale of its officers and men.

In a signed letter, the Prime Minister stated:

Dear Air Chief Marshal Latif,

As you lay down office, I should like to convey to you the Government's as well as my own appreciation of the manner in which you have headed our Air Force, raising the morale of your officers and men and improving their conditions of service. You have brought to your duties high competence, dedication and qualities of leadership and have won the esteem of your colleagues. You have reason to look back with satisfaction on your career.

For the coming years, I send my good wishes to you and Mrs. Latif who took such personal interest in the welfare of the wives of Air Force personnel.

SIGNED

(INDIRA GANDHI)

CHAPTER 02 Some of Idris' Distinguished Ancestors

The Latifs

Adamji Jivabhai was Idris' paternal ancestor, but was also an ancestor of Idris' maternal grandfather Jaffar Hydari. Adamji's son Abdul Latif was born in 1780.

Adamji Jivabhai Latif

The family traded with China, especially Shanghai, Canton and Hong Kong from around the early 1700s. Sir Akbar and Jaffar Hydari's father Nazar Ali Hydari also traded with China, with terrible risks from piracy at sea. Adamji Jivabhai's son Abdul Latif and their descendants continued as wealthy and successful merchants in the China trade up until the early 19th century. One or two of the sons or nephews would go in turn and live in China for three to four years at a time. They were from Cambay, which was a port of some importance in coastal trade, with strong commercial and cultural links with Yemen right up to the early 20th century. From Cambay, they moved to Mumbai.

Camruddin Latif, Amiruddin Latif, Vazirunissa Latif seated
Camruddin's two daughters and son Alma standing

The family belongs to the small Sulaimani Bohra community, a subsect of the Ismaili Shia community, largely based in Gujarat. They were converted to Islam in earlier centuries by missionaries from Yemen. There was some admixture of Arab blood, and the links with Arabia continued to be manifest in the marked use of Arab words used both in Gujarati and the Urdu that the Sulaimanis adopted in the mid-19th century. By now, so many members of the family have married those from other religions and races from all over India — and also from different countries in many parts of the world — that they are a cosmopolitan group.

Idris' father, Hasan Latif and mother, Leila

Idris' father: Hasan Latif

Idris' parents: Hasan and Leila

Hasan was born on 2nd October 1882. He was a tall, big man who had done his schooling at St. Xavier's in Bombay. He was very good at his studies, and received First Prizes in English. He also knew Urdu and Gujarati and, in fact, always preferred to do mathematical calculations in the latter language. He later went to King's College, Oxford, graduated in engineering and became a member of the Institute of Engineers. He then went to Heidelberg in Germany to finalise his study and specialisation in Civil

Engineering. After returning from Germany, he was employed as Assistant Engineer in the engineering department of the Nizam's government, where he later became Chief Engineer of the state. This entailed a great deal of touring throughout the state, and as Hyderabad state was about the size of France, it involved a lot of travelling.

He was so meticulous in checking on even the smallest details in the implementation of the work, that he would go on horseback to areas where roads were being laid, and personally check even the small stones, called 'metal', to be laid for the roads. Each had to be cut to a required size. Those roads of Hyderabad state were known to be excellent, and many have lasted and are in good condition even now, after almost three quarters of a century. Today, most of our roads are recently laid but are full of potholes, even after one monsoon season. Many of the fine government institutions and buildings in Hyderabad were also set up during his tenure as Chief Engineer of the state. After he retired, he became the Principal of the Osmania Engineering College and then went to Udaipur as he was invited to be the Chief Engineer there.

After his marriage to Leila Hydari, they lived for a while in the district of Warangal. Their home was a huge mansion, with long verandahs edged with arches over columns. He was referred to as 'Bajani' by his children, an abbreviated form of 'Bawa Jaani' or 'dearest father.' He was a very caring, family man who loved to have his dear ones near him. They went on many outings, riding on horseback, or going on picnics together whenever it was possible, and he loved to have friends over. He and Leila spent hours together in their garden, growing their lovely plants. He had a beautiful home at Pali Hill in Mumbai, and lived there for several years after his retirement. When he sold it later, it was bought by Dilip Kumar, the great actor. Bajani was a strict disciplinarian and a stickler for punctuality. It was almost that if you were not at the dining table at the times fixed for meals, you may have to go without that meal! The children, thus, learnt to be punctual and their digestive systems must have benefited by this. It was also stressed that it is very inconsiderate and rude to keep anyone waiting, no matter how low or high their status.

Hasan was a diabetic for over 30 years and gave himself the required insulin injections before meals, twice a day. He never fussed about it, enjoyed meeting friends, whom he met with great warmth and affection. Hasan and Leila both had a well respected homoeopath as their doctor, and remained active and healthy for most of their lives. Even when he

was over 70 years old he was so full of energy, that occasionally while they were living in Hyderabad at their home 'Latifia', he would suddenly decide to go on a visit to their home in Bombay, and they would pack and drive off by road to Bombay in their car on the following day. For many years, he always had a Ford car, replacing it with the latest model every three years or so — but always a Ford. While all the children were home, they had two Ford cars for all of them to fit in. His habit of reading to the children from religious books, both of Islam and other faiths, every morning at 7:30 am, instilled in them his own respect and knowledge of the great scriptures and faiths of the world, in addition to their own faith, Islam.

Idris' mother Leila

Idris' mother, Leila

Leila's mother, Surayya

Among the most lovable of the women of the family, and one of the finest human beings to emulate, was Idris' mother Leila. She was born on 4th November 1898 as the eldest of three children to Surayya nee Lukmani and Jaffar Hydari. Both her parents died when she was just a little child. Fortunately, her father's brother Sir Akbar Hydari, and his wife Amina nee Tyabji, took over the three children and brought them

up as their own. They grew up along with the children of the Akbar Hydaris as close brothers and sisters, with the same fine principles.

There are many books which were awarded to Leila as prizes for doing brilliantly during her schooling at the Mahbubia Girls School, the most elite school of Hyderabad of that period. Leila was married to Hasan Latif, and though there was a large gap in their ages, they developed a most compatible and happy relationship. Leila was a wonderful mother, wife and homemaker; Hasan, a loving father, and both were very family-minded. Whoever met Leila was charmed by her. Not only was she lovely as a person, but she was quick witted, with a sense of humour, and a lady of sincere compassion. She became Commissioner for the Girl Guides for Hyderabad and was involved with several social causes. She also saw to it that the children of her own household staff went to school, which was not always the norm at the time. Once when she was told that the son of her woman garden assistant was bunking school, she had him fetched in her car and taken to school! She would never tell an untruth, would not gossip or speak ill about others. She tried not to hurt anyone and helped many who were in need. In her later years, a blind girl who had joined college would come every day and Leila would read out her college textbooks to her and help her with her lessons. The girl went on to graduate and later became a government high school teacher and continued to teach there till she retired.

Leila and Hasan were passionately fond of gardening and spent much of their free time looking after their garden. Leila also received numerous prizes when she entered flower pots in the flower shows, which were written about in the newspapers of the day.

Hasan died in April 1957 at the age of 75 after a sudden heart attack. Leila insisted on living in their home on her own thereafter. She prayed regularly, having her meals at fixed hours, visiting friends in the evening. Leila would be working in the garden at 7:30 every morning till it was time for breakfast, and then again in the early evening. She remained healthy and led an active life for many years. Travelling on her own, she regularly visited the children wherever they were. Whenever we came to spend our holidays with her, she would load us with delicious homemade goodies, often made from her garden fruit. At the age of 86, she gradually became bedridden for almost four years with Alzheimer's disease and passed away on 28th July 1989, when she was almost 91.

SOME OTHER ANCESTORS OF IDRIS' MOTHER, ALSO RELATED TO HIS FATHER

Badruddin Tyabji Sr

There have been numerous ancestors and relatives of Idris, both men and women, who have distinguished themselves in different fields. The majority have been those who worked with great dedication for the betterment of the underprivileged and those who were deeply involved in our freedom movement, and in the service of the state. This ethos of integrity, of patriotism and of always doing what is right has probably influenced Idris' upbringing. As early as the late 19th century, there were special examples of this through Badruddin Tyabji, Camruddin Tyabji and Amiruddin Tyabji. The three brothers, at the time of the great Liberal Movement from the 1880s onwards, were involved in institution building, the judicial system, the educational system, the Municipal Corporation, the Legislative Council, and so on. Badruddin Tyabji was the first Indian barrister, and his brother Camruddin, the first Indian solicitor. Their sister was Idris' great grandmother, the mother of Idris' maternal grandfather Jaffar Hydari.

Justice Badruddin Tyabji was the first Indian judge on the original side of the High Court, and he was elected as the President of the third Indian National Congress. Badruddin Tyabji (derived from Tyab Ali) was born in 1844. His father was the scion of an old Cambay emigrant Arab family. After passing the London matriculation, he joined the Middle Temple, became a barrister (April 1867) — the first Indian barrister in Bombay. In 1895, he accepted a judgeship of the Bombay High Court. This, as Sir Pherozeshah Mehta stated, was as enthusiastically welcomed by the Hindus and the Parsis as by the Muslims. He acted as Chief Justice in 1902, the first Indian to hold this post in Bombay. He was known as a great judge and for his courage and impartiality. This was typically shown by his granting bail to Tilak in a sensational case after it had been rejected thrice by others, and by admonishing eminent British counsels for denigrating the Indian National Congress and Indian character. He is said to have stated, "I have always regarded it (Congress Presidentship) as the highest honour, and higher than being on this Bench ... let me tell the counsel that in my Court no contemptuous reference to that body will be permitted."

Badruddin Tyabji Sr

From July 1871, Badruddin was prominent in the agitation for an elective Bombay Municipal Corporation, and topped the list of those subsequently elected on that body. Badruddin Tyabji, Pherozeshah Mehta and Kashinath Telang were popularly known (in that order) as "The Triumvirate", "The Three Stars" of Bombay's public life. In 1885, he helped to found the Bombay Presidency Association and virtually ran it all by himself. Badrudddin had denounced all communal and sectarian prejudices. He missed the second session of the Indian National Congress due to ill health, but was unanimously elected President of its third session in Madras (1887).

Badruddin's own education and background were a harmonious blend of the East and West. Camruddin and he were principally responsible for establishing the Anjuman-i-Islam in Bombay (1876) for the betterment and upliftment of Mussalmans in every direction. Badruddin was acutely conscious of this need, particularly among the Muslims. He campaigned against 'purdah' all his life, holding that it went far beyond the Quranic injunctions. His was the first Muslim family to discard it, and his daughters were the first to be sent abroad for education. He supported the Age of Consent Bill (1891), despite Hindu and Muslim opposition.

Badruddin's all pervasive intellectual and personal distinction enabled him to exert considerable influence for worthy causes on the more enlightened Englishmen, without loss of personal or national dignity; in fact, with an accretion to both. He was not only, as Mahatma Gandhi wrote, "for years, a decisive factor in the deliberations of the Congress," but one of its creators. It gained its national character by Muslim participation largely through his influence. On Badruddin fell the main burden of counteracting the Two-Nation theory. This he did without abating a jot of his zeal for the advancement of the backward Muslims. Among Muslims, Badruddin was the first

to create a secular political consciousness; and, nationally, he was a pioneer in making it the Indian ideal.

(The above includes excerpts from a speech by his grandson, the former Ambassador Badr-ud-Din Tyabji).

AKBAR NAZARALLY HYDARI — PRIME MINISTER OF THE DOMINIONS OF HYDERABAD, 1937-1940

Idris' maternal grandfather was Jafar Hydari, the brother of Sir Akbar Hydari, the former Prime Minister of Hyderabad state. Jafar and his wife Surayya died young. Their three little children — Leila (Idris' mother), Tyaba, and son Rafi Hydari — were brought up by Sir Akbar and Lady Hydari as their own.

Sir Akbar Nazarally Hydari, Nawab Hyder Nawaz Jung

Sir Akbar was born in Bombay in 1869 and died in 1941. He was named Mohammad Akbar Nazarally Hydari, and was the nephew of Badruddin Tyabji (sister's son). He was of a family of Oriental and Western learning, and of integrity and intelligence.

He had studied English and Latin at St. Xavier's College, Bombay, and had acquired a B.A. degree at the very early age of 17, before passing the competitive examinations with distinction. In 1905, he was invited by the state government to be Accountant General, later to become Financial Secretary, then Secretary in various other departments, then Finance Minister. Earlier, he had refused a post when a British officer was promoted over him in 1937. He was made President of the Executive Council and Prime Minister of the state from 1937 to 1940. He was responsible for the financial revival of the state.

When Sir Akbar became Prime Minister, the state was in a financially straitened condition, and required a large amount of funds for progress to be initiated. With the Nizam being a benevolent, dynamically supportive and caring ruler, and with Akbar Hydari as his Finance Minister and later Prime Minister, they were able to bring about substantial surplus finances for the state, with honesty and ability. This was in spite of making provisions for debt redemption, post-war development, public health, education, vast irrigation projects, railways, road and air transport, contributions to war efforts and for famine insurance. As the farmers suffered terribly in times of drought, the government began paying compensation to farmers when crops were ruined. Hyderabad also had the special achievement of being the lowest taxed state in the world! The Civil Service was also revived, making it open to all, with entrance examinations based on merit. With the British, he insisted on the state having its own coinage, the Hali Sicca, which was also minted in the state, and was almost on par with the rupee of British India.

The Hyderabad State Bank was established with branches in the districts. Air links were brought about when the state set up the Deccan Airways along with training facilities for pilots as well as a flying club. The Nizam's State Railway became known as the best in the country. Industrial development was taken up and the Godavari Valley Development Authority for Hydroelectric Power was formed to augment the irrigation of 2,700.000 acres of land. In 1941, the Industrial Research Board was set up, which included the development of handicrafts and small-scale industries. The Singareni Collieries, cotton mills in three districts, paper, iron, steel and cement factories were established as well as gold mines at Hathi and Kollur. An Industrial Trust Fund was created at the time for financial help for industry. The Archaeological Department was started and the Ajanta and Ellora Temple caves, which are now World Heritage

sites, were excavated and painstakingly restored, sparing no effort or expense to do so. Their conservation and restoration were undertaken after identifying and getting experts from Italy, and a vast amount of funds was allocated for this. Sir Akbar was a man of great ability and efficiency, and served loyally under a most forward looking ruler, the VIIth Nizam, Mir Osman Ali Khan.

As the delegate from Hyderabad at the Round Table Conference in London, he worked for his ideal of a united India. The restoration of Berar to Hyderabad from the British was of great concern. In 1930, this was taken up at the Round Table Conference where Sir Akbar Hydari represented Hyderabad, and raised the issue of Berar. Finally on 24th October 1936, an agreement was reached by which the sovereignty of the Nizam over Berar was accepted. Earlier, he had been knighted by the King, and had also been given the title of Nawab Hyder Nawaz Jung by the Nizam. Later, he was given the Oxford D.C.L. and the French government bestowed on him the title of Grand Officier D' Aragon D' Annam.

From *Time* magazine dated Monday, 22nd February 1937, an article on the 7th Nizam refers to Hyderabad and "its key statesman and Minister, the Nawab Sir Akbar Nazarally Hydari, eulogising his achievements. Sir Akbar was a devout Muslim, of secular outlook, a man who welcomed people of all spiritual learnings, of all faiths and beliefs, as well as artists, dancers, musicians and writers, and they flocked to Hyderabad as well as his home from all over India and from abroad. When he finally moved to Delhi in 1940, it was with sadness and reluctance, but the Nizam had agreed to the request from the British government for Sir Akbar to join the Viceroy's Council. He died there within a year in 1941. In a public condolence, His Exalted Highness the Nizam highlighted Sir Akbar's contributions in numerous spheres."

Sir Akbar always gave a great deal in charity, but did not own any property himself. When he passed away, his collection of Indian miniature paintings was donated according to his wish, to the state museum and to the Prince of Wales Museum in Bombay. He was a great personal friend of Mahatma Gandhi and used to drive him around Hyderabad himself, when Gandhiji visited Hyderabad.

LADY AMINA HYDARI

Lady Amina Hydari nee Tyabji

LADY AMINA HYDARI was also a niece of Badruddin Tyabji. She did a lot of' work for the poor and for the betterment of women. She was actively involved in helping victims of the floods even when she was obviously in an advanced state of pregnancy. Her father was from Yemen, a holy, learned and senior member of their Suleimani community. She was also known for the elegance of her home and for her personal dress sense. She was actively involved in setting up the Mahbubia Girls School along with Sarojini Naidu and my own paternal grandmother, Tyaba Begum, and two others, as well as in other welfare works, especially for women and children.

ABBAS TYABJI ... a true Gandhian

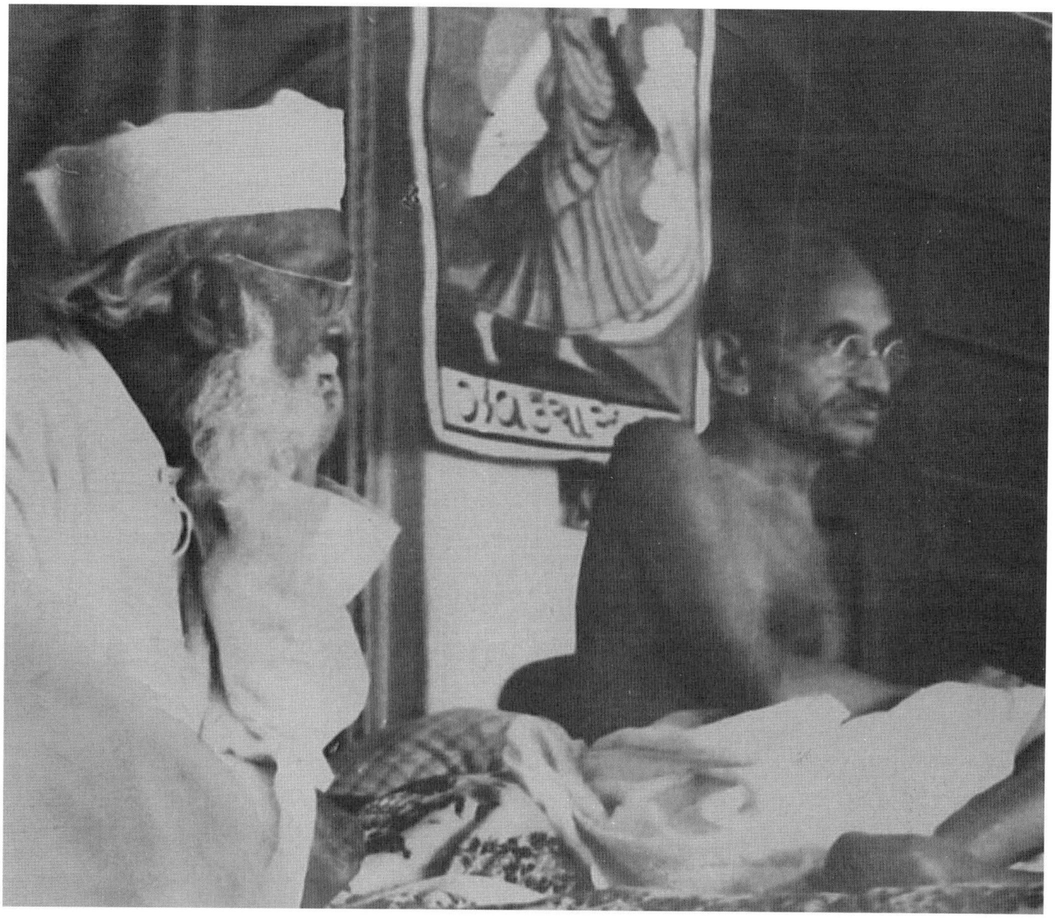

Abbas Tyabji with Gandhiji

Abbas Tyabji, who was the son in-law of Badruddin Tyabji, is a great name in the freedom movement, and was a dear friend of Gandhiji. He resigned from the post of Chief Justice of Baroda state when, appointed on the commission to enquire into the Jallianwalla Bagh incident, he discovered the enormities that were being committed by the British rulers and forthwith joined the Independence movement. When Gandhiji was expecting to be jailed in May 1930, he nominated Abbas Tyabji to act in his place and lead the Dandi March. However, Gandhiji led the Dandi March himself, and then he was arrested. Abbas Tyabji, as nominated by Gandhiji, then led the *satyagrahis* to

the salt depot at Dharsana and the country resounded with the slogan, *Khara rupaya chandi ka, and raj Tyab-Gandhi ka*. Salah Tyabji, the son of Abbas Tyabji, was the leader of the Congress in Rangoon. The family described to me how all foreign clothes from the U.K were piled up outside their house and burnt when he became a follower of Gandhiji.

Others who participated actively and were jailed during the movement, sometimes repeatedly, were Jabir Ali (who was offered the post of Agriculture Minister in the first government of the state of Bombay, by his great friend, Chief Minister B.G. Kher), Farid Tyabji (with considerable involvement in village reconstruction and removal of untouchability in the Ahmednagar District), and Saad Ali, actively involved in the freedom movement and jailed for it, and also involved in social reconstruction and other causes.

Justice Faiz Tyabji of the Bombay High Court, Justice Hatim Tyabji, Justice Husain Tyabji and Mr. Danial Latifi are the most prominent in the judicial field.

For environment and wildlife are the well known ornithologist Dr Salim Ali, as well as Mr Zafar Futehally, both of whom were recognised with National Padma Awards. Another was Mr. Humayun Abdul Ali.

In the field of medicine, we have had Dr. Budruddin Lukmani, FRCS, MRCP. For distinguished careers in the ICS, we have Dr. Alma Latifi, CIE, OBE, Financial Commissioner, Revenue of the Punjab, and Mr Hamid Ali as well as Mr. B.F.H.B. Tyabji, Ambassador to West Germany, Japan, Iran and Belgium, Vice-Chancellor of Aligarh Muslim University.

The first Vice-Chancellor of Kashmir University, and then Ambassador to Egypt, was the legal luminary, Professor A.A.A. Fyzee. His sister was Nazli Begum, married to the Nawab of Janjira. She was a well-known poetess. A woman writer was Asif Fyzee's wife Mrs. Sultana A.A.A. Fyzee, whose creative writings in Urdu are prescribed for study at universities.

The eminent historian, Professor Irfan Habib, belongs to the family, for his father, Professor Mohammed Habib, married Sohaila Tyabji, the daughter of Abbas Tyabji.

Several women of the family were remarkably active, and many of them achieved

considerable prominence in areas relating to women and children, in village and social reconstruction, and in demonstrations against colonial rule, leading to arrest and imprisonment. Some of them were known independently for their own political life, some for their participation in the political life of their husbands, and others for their contribution to education, several more for social welfare and integrationist activities. Cultured and well-educated, some of them having been sent to England for their education, they stood out not only among Muslim women but among Indian women as such. But even more important was a notion deeply imbued in them of service to society and the nation at large. Several women of the family were also actively involved in the movement through picketing, and various other types of protest, often against alcohol also. Two eminent ones were arrested and jailed for, thus, picketing, but were soon let off, possibly because of the public outcry as expressed in the newspaper clipping given below from the *Bombay Chronicle* of 10th July 1930.

HARD LABOUR FOR MRS. LUKMANI
AN OUTRAGE 'ON INDIAN
PUBLIC OPINION.'

BY G.K.N.

Savage and unconscionable. This must be the public verdict on the sentence which the resident First-class magistrate of Kurla thought fit to pass in the case of Mrs Sakina Lukmani who has been sentenced to four months rigorous imprisonment. I do hope there is some error in the reporting of the case. Possibly the treatment to political prisoners, even when officially styled rigorous, is directed to be not too harsh. Yet to condemn to prison like a common felon, a person of the convictions and culture of a daughter of the late Justice Badruddin Tyabji is an outrage on Indian public opinion. Honest and avowed Moderates, such as I pride myself to be, have borne with patience the excesses of the Extremists. Are they to be rewarded with this eruption of the Governmental excesses through their myrmidons as expressed in the judicial atrocity perpetuated against Mrs Lukmani? I am against the picketing of liquor shops which are certainly not in the spirit of the religions for which so much has been made. But comparing the savageness of the sentence with the offence of the lady, the latter appears to

be a trifling peccadillo. Four months rigorous imprisonment appears to me to prove the sentencing Magistrate in a condition when the public must demand a certificate of his suitability for the post before he is permitted further to function on the Bench.

BOMBAY CHRONICLE, 10-7-1930

Two of Abbas Tyabji's daughters also played prominent roles in the freedom movement, one of them, Raihana Tyabji, through actual political involvement, and the other, Sharifa Hamid Ali, through a very full career of social work, particularly in the upliftment of women. Raihana Tyabji was a very prominent Gandhian. Referred to by the Mahatma as "my beloved daughter", she devoted her life to the freedom struggle and to life in his ashram.

Mrs. Fatima Tyabji, along with her husband, a grandson of Badruddin Tyabji's, abandoned a comfortable life in Bombay, on the call of Gandhiji to live in a village, take up farming and village reconstruction activities, especially the cause of the Harijans. They chose the Ahmednagar District where they lived thereafter and succeeded in doing this.

CHAPTER 03

Stepping Stones: School, College and into the Air Force

Idris Latif, 1940

As I have mentioned earlier, my husband Air Chief Marshal Latif has always shied away from publicity, and has very seldom agreed to be interviewed. However, in 2009, a gentleman persuaded him to answer a questionnaire as he said that he was producing features for television on the Defence Services, and he had nothing on the Air Force. The Air Force is so close to Idris' whole being that he agreed to respond only because, I think, it would give the Air Force a place in the sun! The questionnaire also had several queries of personal and family concern to which Idris responded by writing them out, but then withdrew them from the televised version. As they provide a brief background, I am reproducing some of them here in his words:

To give you a bit of ancient history, my ancestors, originally from Arabia, migrated to India almost a thousand years ago; initially came to Gujarat

but now for several hundred years have largely settled in Maharashtra, primarily Mumbai. It is with some pride that I can say that, a number of our families were close to Mahatma Gandhi and very actively associated with the Freedom Movement. My earliest years were spent in Warangal where my father was stationed for just a few years.

Idris (centre, front row) with his parents and six siblings. Warangal, mid 1920s.

Idris' first family home...early 1920s

Later, we moved to Hyderabad, and right through my school and college, we lived in a lovely spacious house in Khairatabad, all nine of us, seven siblings with our parents.

Idris' first home after moving to Hyderabad; late '20s to end '30s

What memorable years they were! What closeness; what harmony! To begin with, the understanding of their individual roles between my parents was, truly, so heartwarming. Another exemplary aspect: never, ever was there a harsh word or serious disagreement between them — at least never, ever within the hearing of any of the children. We tried, perhaps, to learn from them, sadly not always successfully! The togetherness of the family, the consistent spirit of mutual accommodation, the wonderful understanding amongst parents and brothers and sisters, was a source of both strength and unadulterated happiness.

But Idris, more than once, would add with a smile,

You ask me about something of my early years with my parents; and I spoke of our wonderful parents and siblings — no less than six of them! — and, the very wonderful memories of my years with them. But I have something

to add, something no less important in my life; perhaps, in fact, even more after I left home to join the Air Force, what was to be my next home. This was and is, truly, my life, with you as my wife. And with sincerity I say that of all our years together, I have such very cherished memories that are no less wonderful. Always, how loving, how caring you have been, and always, always so very, very selfless. In fact, life with you, believe me, has left me with a treasure of memories by no means and in no way, less precious. And in just a few years after our marriage, with our Mariam and our Asad and our Asgar; each again, also, so very loving, so very caring — all five of us together. Further in time, our life with our Sandy and our Fozia and our Huma — each of them again, each in his or her own way, so affectionate, so caring. And, finally, now, our darling grandchildren, each not only very good looking, but again, without exception, so well brought up, so loving, and with so much courage and character. God couldn't have blessed us more. Could we really have ever, ever have asked for more? Even dreamt of more? God must truly be thanked for all His many, many blessings. And I have to tell you that, over the years, it is, at least partly, why, in gratitude and thanksgiving that I pray, every day, five times a day.

Knowing that Idris was one of the best students, always in first or second place in his class at school through the years, I asked him, to share with me something about his school years. In the very early years, he was at St. George's Grammar School, when he and his younger brother were perhaps just six or seven years old. As Idris said, "My later schooling at Madarsa-e-Aliya and my years at Nizam College — both fine institutions — taught me much, but in terms of education, in the academic sense. In a broader sense of intellect, character and personality, it was, without doubt, the home environment that was the most valuable and enduring part of my education. This included the moulding of my character, my sense of values, the importance of honesty, of integrity, of respect for elders — rich or poor, including the household staff. The importance of punctuality too, which is, in essence, consideration for others. The importance, particularly in our rich multi-religious society, of respect for, and a true understanding of all religions, an essential feature being that all are one — all are God's creation — again, in essence, a guide to one's own conduct:

Khuda ke ashik to hain hazaron, banon mein phirte hain mare mare;

Main uska banda banunga jisko khuda ke bandon se pyar hoga.

Idris adds that, of Madrasa-e-Aliya and, again, of Nizam College, he can recall little, very little, of what happened — that is, in class — except for their very unusual type of Urdu classes in school. Their Urdu teacher, Agha Hyder, was a very lovable and learned man, a distinguished descendant of a Moghul grandee. He told the boys that it was best not to waste class time on textbooks which, he said, could easily be covered at home. Instead, he almost invariably divided the class into two teams for *baiyt-baazi*, a popular type of Urdu poetry contest. This was the best stimulus, he said, not only to memorise memorable verses but, with it, an excellent opportunity for explanation and understanding. This proved to be so very true; that there are still so many poems or *ashaar*, that Idris can, with ease, repeat from memory — and often does. Agha Hyder taught his students all the nuances of each poem so that they came alive for them. There is now a museum section dedicated to him in the home of his distinguished son-in-law and daughter Nawab and Begum Moazzam Hussain.

In spite of Idris' statement that he remembers little or nothing of the classroom periods of his schooling, he must have paid some attention to his class work at the time, as is evident by the many books that he was awarded as prizes for coming first or second in his class. The number of awards accumulated over the years proved that, in spite of times of fun, he always did extremely well in his studies. Urdu, he once confessed, occasionally pulled him down. However, thanks to Agha Hyder, he learnt to truly appreciate and enjoy the finer shades of expression and the depth and beauty of Urdu poetry.

Another wonderful feature of the years in school and college that he often recalls was that there was little or no need for homework, which was often undertaken and completed in just an hour after dinner. During *Ramzaan*, he did his homework before the *sehri* breakfast and prayers early in the mornings, well before dawn, or just after the *sehri*. Ninety percent of the free time at home was for fun with the family, cricket, tennis, badminton and a game called 'I-spy,' as well as kite flying during the season, which was a hot favourite. A common activity among the children was testing their skills with the catapult, especially popular in the case of the two youngest, Nazir with Idris. Come December and January, kite flying had top priority.

Their daily routine, Idris recalled, was to get up in the morning at around 6:00 am and get dressed by 6:45 am for prayers. The prayers culminated with his father's readings of a passage from the English translation of the Quran, followed by a few selected short prayers collected and translated also into English; and when the Quran was completed, passages from the holy books of other religions. Breakfast was served punctually at 7:30 am, and then by 8:30 am, Idris and Nazir were on their bicycles and off to school, just 15 minutes away. Each night, from 7:30 pm until dinner time at 8 o'clock, their father read to them from the English and other classics: Dickens, Scott, Alexandre Dumas and others. At home, Idris said when questioned, they usually wore kurta pyjamas, but with a *fez* cap, or *roomi toopi*, and a *sherwani* for school and social gatherings. His father read well, and his reading of the classics to them every day enabled the children to listen to good English which was spoken well, and helped in their own command of the English language.

Saeeda, Idris' sister in a classical dance pose, 1941

Idris with his first and only panther, 1938

Idris once mentioned an episode which, he added, may sound amusing today but proved 'painful' at that time — and painful in more ways than one! Nazir and he, having done particularly well in their exams, were told that they could each choose a new bicycle. A good bicycle, he said, cost about Rs. 50 at the time. Nazir got a Raleigh, and Idris selected a Philips; each, understandably, claimed that his own bicycle was the better one. The best way to settle the dispute and determine which was superior, they decided, would be to test the quality of their bicycles by riding them and colliding into each other. The outcome was, of course, the considerable damage done to both their brand new cycles. It resulted also, deservedly, he even now confesses, in a good punishment from their father!

All of the children, all seven of the brothers and sisters, loved horse riding and for several years, even when in school, they were given the benefit of good lessons at a training institution of the state forces at Masab Talaab. Often, on holidays, all the children would go riding towards, and occasionally even within, the Golconda Fort, where chasing deer was usually the favourite climax.

L to R: Leila, Hasan, Mahboob, Idris (at centre), Nazir, Saeeda, Akbar, late 1930s

Yet another wonderful opportunity to practise riding was their annual summer holiday. Most years, Idris mentioned, their father would arrange his inspection tour of the districts from late April to the end of May. The whole family would set off in their two Ford cars on these truly memorable trips: partridge and deer shooting, with wayside breakfasts or picnics en route. They had comfortable stops for a couple of days at selected Dak Bungalows, or Government Rest Houses. Very important for the children, was that a pony for each would be waiting at every stop. Idris continued to love horse riding and, even years later, when he was at the Joint Services Wing and then at the Defence Services Staff College at Conoor where horse riding was possible, he was a regular participant in the weekend hunts. He even participated in the Ooty Gymkhana Gentleman's Race, only coming third, he confessed, but recalling that it was great fun. Other favourite games were cricket and tennis.

Standing (L to R): Idris, Akbar, the bridegroom Badruddin, Mahboob, Saif, Nazir
Seated (L to R): Saeeda, Tehzeeb, the bride Surayya, Hasan, Leila, Vazir, 1939

After Idris finished the matriculation, he joined the Nizam College. Because of his great love of forests and wildlife, his ambition, he told me, was to join the Hyderabad Forest Service. However, when he was in college, just 17 years old, and World War II had started in Europe, he saw very attractive posters encouraging people to join the Air Force. What an exciting and challenging career it would be, he thought. And the candle was lit!

When he first spoke with his parents about joining the Air Force, they had not been happy about it. In fact, Idris told me not too long ago, that his mother had tears in her eyes, and seeing her reaction, he said, he couldn't dream of pressing the thought. However, when he again broached the subject some six months later, they readily agreed, saying that in his future life they did not want him to feel that they had stood in the way of a career that he wanted. As Idris continued, "What parents! What a lesson for me for the future. I can never forget the understanding and unqualified support that I received from each of them for ever thereafter."

CHAPTER 04

Into the Cockpit

The Air Force was then called the Indian Air Force Volunteer Reserve, later to become the Royal Indian Air Force and then, finally, on Independence, the Indian Air Force. To join the Service, Idris had to be interviewed by the British Resident at the Residency in Bolaram, which is now the Rashtrapati Nilayam in Hyderabad. When he arrived (well in time, as always!) there were, he told me, about 12 boys sitting in a line, waiting to be interviewed. Most of the others wore Western clothes but Idris went in a *sherwani* (*achkan*) with a *fez* cap, as was worn in Hyderabad on formal occasions.

Idris, 1942

When questioned by the Resident on why he wanted to join the Air Force, his unhesitating response was that he felt that it could, particularly with the war on, be full of excitement, and later also, as a career,

both challenging and emotionally satisfying. It was a brief interview that day, and he later found that he was the only one selected from the group in Hyderabad. The next step, after the interview, was the medical check-up which, he was relieved to find, proved to be generally satisfactory. However, a cause for some initial worry was when he was told that, because his father had diabetes, he would have to stay overnight at the Military Hospital for further blood tests. The evening and night were fraught with growing apprehension about a possible setback, in fact, rejection, if the tests were to indicate an excess of blood sugar. The next morning he was marched to the medical investigation room where a Major poked his arm painfully several times with the needle, unable to locate the vein, and finally apologised, claiming he was "a little out of practice!" However, when he finally succeeded in drawing the blood and it was tested, fortunately, Idris was cleared to join the Air Force. While he was thrilled and excited, he soon sensed that, although his mother bravely joined in the celebration, in her heart was deep concern about his future of flying, and, of course, the inherent dangers of flying military aircraft. But now this was compounded, particularly with the war on as well. Yet, she never spoke about any of this because she knew how eager and excited her son was. What a father and a mother — one in a million — as Idris always said.

After joining, Idris first went to the Madras Flying Club to complete 25 hours of flying on the Tiger Moths. Later, he learnt that this was to test one's flying aptitude before being finally put in uniform. Idris soon made some very good friends during his three to four weeks in Madras. Ultimately, however, apart from himself, he said, sadly only four others passed the tests in Madras. The Chief Instructor at the Flying Club was Tyndal Biscow, mature and understanding, as well as a fine pilot. So also was Idris' instructor, Dastur; Idris often told me he was "altogether, in spite of an aggressive looking hooked nose, a very fine man!" As Idris explained, "If you are fortunate enough to have a good instructor, it makes all the difference." And if Idris learnt a little more quickly and was able to do an early solo, he believed it was because of Dastur. For the final clearance, Flying Officer Pitt Browne, a senior instructor at the Initial Flying Training Wing of the Air Force at Begumpet, came for the flying check before the formal induction into the Indian Air Force Volunteer Reserve.

Idris and the few others who were selected were then sent to the Initial Training Wing at Walton, in the vicinity of Lahore. He was one of the no less than 65 others who were formally inducted as Officer Cadets of the 11th Course. It is shattering, Idris once reminisced, that of the 65 young Cadet Officers, who included Dani, Damo Oak, Ramakrishna, John Mckenzie and so many, many others who became close friends, almost all of them had passed away by the time that Idris was Air Officer Central Air Command in 1974. To Idris, the hours of learning to fly in the Air Force were incredibly exciting. However, as he soon learnt, this pleasure was to come only later, once they had completed the six to eight weeks of training at the Initial Ground Training School at Walton. He soon found that the focus was all on physical training, training on the parade ground and in the classrooms.

In all this, what Idris found very difficult to accept and deal with was the vulgar language used by many of the British team members: foul swear words punctuating almost every sentence. This was something to which he had never ever been exposed. In fact, he was so horrified that, after a couple of weeks, he seriously considered resigning. However, one of Idris' colleagues at that time was Ramakrishna, who was with them for training and to whom Idris had taken quite a liking. Ramakrishna had already served in the Air Force for several years and was now applying for a commission. He cautioned Idris not to act so impulsively: "I have been in the Air Force a long time and you will find that there are many, many others — in fact, most others — who are just wonderful people with their hearts in the right place." Idris stayed on and, to his relief, in later years, found that Rama was dead right!

Idris had his first taste of Air Force discipline very soon after joining it. At this stage, he recalled, he had very little facial hair — only a slight growth on his chin, and, therefore, he shaved only every few days. Thus, it was not long before he was called up and sternly reprimanded: "Latif, in the Air Force, we shave every day." Another aspect of the very strict discipline imposed was that the cadets were not permitted to leave camp until they were able to march and salute correctly. This seemed to take an eternity — certainly a good month of the total 18 weeks of their training at Walton! It was only during their last week that they were able to go off the camp and into Lahore to sample the marvelous chicken curry and excellent *warquii* parathas about which

they had heard so much. As a cadet in those days, Idris recalled, his first salary was 125 rupees: This was enough to cover his Mess bill of around 30-40 rupees and to pay a further 30-40 rupees towards his share of the salary of a valet-cum-butler whom he shared with two others.

From Walton, Lahore, Idris was posted to the Elementary Flying Training School at Begumpet, where he had a very able and caring Instructor, Flying Officer Fletcher. The flying was done on the Tiger Moth, an excellent bi-plane trainer of that period, which Idris had flown before, at the Civil Flying Club in Madras, when he had been assessed for his flying aptitude. At Begumpet, first the thrill and excitement of the first solo flight, followed by steep turns and steep dives, soon followed by the excitement and exhilaration of doing aerobatics was enough, Idris said, to convince him that he had selected the right career! It was during this period of training at Begumpet at No. 1 Flying Training School (the other training period being at Jodhpur) that Idris won his very first trophy in flying. This was in the Forced Landing Competition at Hakimpet which, at that time (in 1941), was just a largish patch of open grassland. To ensure that in case of a real emergency, the pilot could make a safe 'forced landing', the instructor would cut the engine at about 1,000 feet and the trainee was expected to bring the plane down really low, down to approximately 10 feet off the ground, to demonstrate that he could glide down correctly without the use of the engine, to make a safe, simulated forced landing. It was only then that the trainee pilot was permitted to power the engine again and resume normal flying! This was not an infrequent test for all trainees and, whilst it may have been quite a challenge at first, after a while, Idris found it quite good fun; in fact, at the Graduation, he found that he had won the prize for (simulated) forced landing!

11th Pilots Course at Elementary Flying Training School at Begumpet, Hyderabad.
Idris in centre row, fourth from left

Idris' first flight at E.F.T.S. Begumpet, Hyderabad, September '41

CHAPTER

05 Fighters at Last

Later, after getting on to Audaxes at Ambala, formation flying was another major step in their flying training which greatly contributed to gaining confidence in the ability to control the aircraft. A little later, low flying, of course, was almost the ultimate in thrills and excitement, especially when posted to a fighter squadron and being trained, in later years, to occasionally flying down to almost tree-top level. When open fields were an invitation, they flew even lower (!) — and then to make the tighter than ever turns, to soar up again. Unfortunately, as Idris said to me some time later, with today's supersonic and very-high-performance aircraft, much of what they were able to do at tree-top heights at that time, sadly, became inadvisable, indeed forbidden, later. After gaining experience on type, flying on instruments (as they have to do when entering clouds) followed by night flying, were also essential parts of their training on fighters. Once again, Idris said, at Ambala, he was blessed with a fine instructor, a Flight Lieutenant Bishop of the Royal Air Force (RAF).

Audax at ATS (Advanced Training School), Ambala

At Ambala, Idris lived in tents, along with

all other cadets/trainee officers. It was freezing cold in winter, but it was exciting to be trained to fly operational aircraft which were then, as far as our Air Force was concerned, mainly the Audax and Hart. On completion of their training, Idris, with the others of the 11th Course, got commissioned on 26th January 1942 as Acting Pilot Officers. Idris was then posted to Karachi to the Sixth Coastal Flight and went on anti-submarine patrols off the coast.

About their total flying training, a remarkable feature during the war years was its short duration. They had barely five months of flying before they were pronounced 'combat ready', against the 3-4 years today in many Air Forces around the world, including our own. Idris joined on 1st September 1941, finished his training by mid-January 1942, was commissioned on 26th January, and went onto operational duties within a week! His first operational assignment was to carry out, as mentioned earlier, anti-submarine patrols out at sea from Karachi with #6 Coastal Flight of our Air Force. They were then equipped with Wapitis and Audaxes, both bi-planes with maximum speeds well under 80 mph, and, sadly, already completely out-of-date. Hurricanes and Spitfires of the Royal Air Force and Messerschmit-109s with the German Luftwaffe were already plentiful and on the front line in Europe.

Wapiti aircraft at Drigh Road, Karachi, 1942

Whilst at Karachi, Idris narrated a trying — and most worrying — time he had to go through: on a trip, on leave, to Hyderabad by train for a few days with his parents, he had a coupe compartment to himself. Before going to sleep at night, he locked the door from inside but kept the windows open. When he woke up in the morning, to his horror, he discovered that his relatively small bag with all his baggage which he had carefully placed under his seat before sleeping, had been stolen. Very fortunately, the uniform which he had taken off at night and had hung in the bathroom, was still there. In those days, each coupe compartment fortunately had a bathroom for its exclusive use. Unfortunately, though, everything else was in the now stolen bag. The most serious loss was of his Service revolver, which at that time, during the war, serving officers were required to carry, and which was also in the now stolen bag with the rest of his belongings. The thief, Idris thought, must have got in through a window, as, in those days there were no security bars provided. To officially report the theft to the police, Idris got down at the very next halt, fortunately being able to wear his saved uniform (instead of the night suit he had slept in!) and made a formal complaint. The police were very courteous but advised him to pursue the matter on arrival in Hyderabad. Of course, Idris never got anything back. But the loss of a Service revolver was a particularly serious offence, especially in war-time, and on his return to Karachi, he had to immediately report it to his Commanding Officer. Within days, he was ordered to go from Karachi to Bangalore, again by train, to face a formal Court of Inquiry.

He had no idea what the result would be, and expected the worst. But the worst was clearly behind him as, to his great relief (and delight!), he got off very lightly. He was punished with only a 'Severe Reprimand' and a fine, believe it or not, of just Rs 35.

After a few months, in 1943, Idris was posted to Kohat where he joined 3 Squadron commanded by Squadron Leader Nanda. Kohat was the first and oldest operational station of the Air Force and when Idris reported for duty, was commanded by Wing Commander Subroto Mukerjee, who later went on to become India's much loved first Indian Air Chief. Kohat was in the Northwest Frontier Province of undivided India. At that time, 3 Squadron still had Audax and Hart bi-planes, great fun to fly but, of course, totally out-of-date. Performance-wise too, they were way behind the fighters already with the Air Forces of the Western countries and Japan.

They were still adequate, in a limited way, for the needs of that theatre, where operational missions were in support of controlling the tribals of the Northwest Frontier Province. Periodically, Idris with a detachment of his squadron flew in their Audaxes to the well-equipped advance base at Miramshah, about 30-40-minutes flight from Kohat. At dusk, all their aircraft used to be pulled inside the high walls of the Miramshah Fort as otherwise, the tribals, not infrequently, fired their rifles from the nearby hillocks and were accurate enough to damage the odd aircraft. All the aircraft with the pilots and ground crews also stayed within the fort from dusk to dawn when, at dawn, they were wheeled out for flying. Idris was given to understand that if any aircraft was, on the odd occasion, for some reason, left out, tribal rifle fire damaged it beyond repair. Idris told me that several of them (very much, including himself) were not proud of these operations against the tribals as the tribals were, in fact, owners of their land and had every right to defend their territory.

L to R: Wing Commander Subroto Mukerjee, Squadron Leader Narendra, Squadron Leader R.H.D. Singh, and Squadron Leader Arjan Singh second from right. Kohat, 1942/43

Whilst in Kohat, Idris, with the squadron, occasionally flew out before dusk for night flying practice to another landing ground in the vicinity, spending the whole night there and returning to Kohat only at dawn. Cooking staff and equipment would also be

sent so that they could all have freshly cooked food and snacks round the clock, each pilot managing to squeeze in two to three sorties. In the morning too, they would be given a hearty, freshly prepared breakfast before flying back to Kohat with their Audaxes. Idris often reminisced that all the pilots greatly looked forward to these occasions which were like picnics with wonderful friends and comrades.

Flying Officer Latif, Kohat, 1942

The morning after night flying, at Kohat's satellite airfield, 1944

To England for Operational Training on Spitfires and Hurricanes

From June 1943 to nearly mid-1944, in anticipation of the modernisation of our own squadrons, Idris, with 24 others, was sent to England. This was for conversion onto more modern fighter aircraft, and also to gain battle experience in World War II against the Germans and Italians. Apart from the thrill and excitement of flying Spitfires and Hurricanes (both of which were amongst the best) with the Allied Forces serving in Britain (then under intense bombardment by the Germans), life in England was, Idris often said, a most heartwarming experience. The warmth with which they were received by the British, the care and concern with which they were looked after, and the excellent training given, has remained for Idris, a truly unforgettable experience. For instance, on the very first morning in London after their arrival, all 25 of them were escorted to a largish hall in which they saw, piled on tables, throughout its length, warm knitware clothing — pullovers, scarves, socks and stockings, gloves, underwear, even balaclavas, all knitted by lady volunteers. “We know that it is summer here, but all of you are from India and we want to ensure that you are well equipped for our much colder climate; so please take all that each of you feel you may need in the months ahead as a welcoming gift from us — and we specially request you not to hold back.”

The next day, Idris and their contingent went on to Harrogate, normally, in times of peace, an exclusive holiday resort in Yorkshire, where they were to await their postings for their flying training. They were, to their delight, all put up in what Idris discovered, was one of the best hotels. To top it all, even though they actually arrived at about 2 in the morning, senior officers were waiting to welcome them and lead them to a magnificent buffet dinner that had been especially arranged to welcome them. Idris often recalls that this was done not only when Britain was at war, and most items of food were strictly rationed, but when the country faced massive nightly air raids across the country by the German Luftwaffe. Really, quite unbelievable; it is no wonder that Idris still talks feelingly of the courage and warmth of their British colleagues. That was not all; a fortnight, or three weeks later, when Idris, with about half the Indian contingent, was posted to their first flying training base at a place called Calveley in Cheshire, they learnt the very next morning that, as a special gesture of welcome, every Sunday, whilst they were stationed there, the Officers’ Mess would be serving Indian food at lunch for all — including those of other nationalities!

L to R: Idris, two friends from the RAF & Flying Officer Biblo Crishna, 1944

Idris at Rednal on a Spitfire

While in England for the final phase, Idris was posted to Rednal in Shropshire, an advanced fighter training base for Spitfires. It was at the height of winter, with everything under a thick blanket of snow. On one occasion, he was briefed to fly in a four-aircraft formation for what was then called a "Rhubarb" practice sortie. This was in essence, at a very low level, at 200-feet above ground level — occasionally, even a little lower — simulating sortie over enemy territory, for effective surprise attacks against opportunity targets. It was mid-winter, the sky was clear; but with the very heavy snowfall overnight, the countryside had become all white, with deep snow. About ten minutes after take-off, Idris smelt something burning. Soon, there was smoke in the cockpit which got pretty bad, and the engine started 'coughing' and, finally, just packed up. Since he was too low to attempt baling out by parachute, he knew that he would have to force-land. He, therefore, pulled up a few hundred feet to get a slightly better view of the surrounding terrain, and a possible area to force-land in. He immediately informed his formation leader that his engine had packed up, and he would be force-landing.

To his delight, he saw an airfield very close on his port (left) and turned in, with the hope that he would now be able to safely glide in and land there. But his hope was very short-lived as he also saw that a formation of fighters had just taken off in the opposite direction. Clearly, Idris had to urgently turn away, trying to stretch his glide as much as possible, and look for an alternative place to land. Soon, he saw one and, as he was already down to 200-300 feet, turned towards it, hoping that under the sheet of snow there were no hidden boulders or ditches. On that relatively short stretch, he was able to put the plane down with its wheels up, skid for a hundred yards or so and stop. Sadly, the Spitfire was damaged but, otherwise, he himself was without a scratch — just a somewhat faster heart beat, Idris confided!

Within 10-15 minutes, an ambulance rolled up and, as Idris still vividly recalls, a particularly good-looking WAAF (Women's Auxiliary Air Force) officer nurse got out, climbed onto the wing and asked him if he was badly hurt. Even though he said that he was absolutely fine, she insisted on feeling his pulse. "A bit fast," she said. "What do you expect?" was Idris' response, "Isn't it normal? A pretty girl like you, holding my hand?" After a stiff brandy at her insistence, he was put into the ambulance and taken to the Mess; and then straight to the tarmac, to fly back to Rednal in another Spitfire.

It was standard policy, Idris told me, that if involved in an air crash, the pilot is made to fly on a similar type of aircraft as soon thereafter as possible. It had been found that, if delayed, some became reluctant thereafter to fly a similar type. The Spitfire, Idris recently mentioned to me, is the aircraft on which he had, till his retirement, the maximum number of hours. Sentimentally, he often adds that it is, or was "a real beauty", and to him, "the most enjoyable fighter I've flown in all my 40 years in the Air Force!"

It was while he was in England that he got to know a very caring and fine English family. He first met them during a weekend holiday when he and a very close Irish colleague, Bill Sloane, were strolling through the rather lovely town of Nantwich, not far from the airfield where he was based. They came upon a house which had an apple tree in the garden which was full of lovely looking red apples. He and his friend decided to ask at the house if the owner would mind selling them a few of the apples. "Nothing like asking!" Bill said, and, opening the wicket gate, went up and introduced himself and Idris, saying that Idris was from India and he was wondering if they could buy just a few. The owners' spontaneous response was that they would be delighted to present them with as many as they wanted; and then most warmly insisted that they come in with them for at least a cup of tea. The tea stretched over a couple of hours, and both Dr. Knowles and his wife – whose home it was – made them feel most welcome, and proceeded to warmly question Idris, believe it or not, about Mahatma Gandhi who, obviously, they greatly admired.

When Idris and his friend got up to leave, they invited Idris to come and spend his Christmas holiday with them and celebrate Christmas with them in their home. Idris, greatly touched, happily accepted their invitation. A few days before the Christmas holidays, however, Dr. Knowles rang up and said that his young son John had chicken pox, adding, however, that as Idris must have had the disease in his own childhood, it should be quite safe for him to come to them. Idris did not remember having had chicken pox, but Dr. Knowles was strongly of the view that as Idris was an adult, he was very unlikely to catch it from John. He added that they both and their two children, Sonia, about seven and John about two, were so looking forward to seeing Idris again, and all of them would be very disappointed if he couldn't come. Idris responded saying that if it really was not inconvenient for them, he would be delighted. Idris, to this

day, vividly recalls the lovely days with them, playing with their two children and the warmth and affection of their parents over Christmas. Sadly, however, when he went back to his unit, he had a severe bout of chicken pox, and was taken to an American hospital which, unlike the British hospitals working under war-time restrictions, had the best of everything, including the kind of food that was no longer available in England during the war.

It was already early 1944, and all the Indian pilots in the batch, including Idris, had been cleared as operationally adequately trained to fly with the Royal Air Force squadrons. Idris was all excited as they were to participate in Operation Overlord, the Allied landings along the French coast to engage and free Western Europe, which was planned for early June 1944. Fate, Idris states, intervened as a signal was received from India that Indian squadrons were very soon to be deployed against the Japanese who had already advanced into Burma, and were advancing towards India. The others of the contingent were flown back without delay. Idris was, unfortunately, still recovering in hospital, but was flown back as soon as he was released — in a Dakota, Idris added with a laugh.

On his return to India, Idris, to his delight, was posted back to 3 Squadron in Kohat, which had already been reequipped with Hurricanes. His Squadron Commander was Squadron Leader Prithipal Singh, whom he greatly admired. Idris often said that his Squadron Commander was not only a fine flyer himself, but was an excellent Commander. He knew full well that his squadron would, very soon, be deployed in Burma, and to ensure that all his pilots, became battle-worthy as quickly as possible, he took extraordinary risks whenever they were flying with him. Once a month, for instance, he would insist that he lead a formation of all 16 aircraft in open battle formation in sections of four and, not only practice rapid cross-over turns, flying at low levels that they all felt were, not infrequently, dangerously low! Then, finally, when returning to base, he would put all aircraft in pretty close echelon starboard formation for the final peel-off for landing. On each such training flight, Idris and quite a few of the others, Idris said, felt, deep down, that at least one of them may not return! To his Squadron Commander's great credit, they always did — and, importantly, with much greater self-confidence.

Another outstanding officer, Jumbo Majumdar, coincidentally had returned at that time from the United Kingdom after operational flying there, and had been awarded a Distinguished Flying Cross (DFC). One day, he asked Squadron Leader Prithipal Singh if he would, after lunch, fly with him and assess if any of the tactics he had learnt in Europe could prove useful. Prithipal, of course, being the man he was, immediately accepted the offer, and, a little later, they went up in two separate aircraft. Tragically, very, very tragically Prithipal was killed in an unfortunate accident during the flight. This was, of course, a most grievous loss to the squadron. To make it worse, this was just a few weeks before No. 3 Squadron was to fly out to Burma. Squadron Leader Shivdev Singh, the Senior Flight Commander, was immediately promoted and given command of the squadron.

Squadron Leader Prithipal Singh, 1944

The Hawker Hurricane

In spite of the tragic event, however, the squadron, now under the command of Squadron Leader Shivdev Singh, flew on the originally scheduled date, 11th January to be exact (as recorded in Idris' log book). All 16 Hurricanes flew in full aircraft formation (with Idris in his favourite Hurricane LE-760 as No. 2 to Minoo Engineer, his Flight Commander) setting course for Burma, accompanied by a Dakota carrying a small maintenance party.

CHAPTER 06

Wings Over Burma

From Kohat, staging through Lahore, Delhi, Kanpur, Allahabad, Gaya, Alipore and, finally, Chittagong and Cox's Bazar, on 16th January 1945, as Idris's log book shows, they arrived at Bawli North, in the Arakan region of Burma where they were based till the end of the war. Bawli Bazaar close by, they soon found, was only a little village of may be a dozen families about five miles away. Bawli North (often so called) was a relatively small airfield with a grass runway, and patches of thick forest in close vicinity. All ranks lived and worked there out of bamboo huts, using bamboo tables and chairs for their technical offices and domestic accommodation, for officers and other ranks alike. They all slept on bamboo beds, working and eating off bamboo tables — and throughout their stay, rather stylishly, even drinking from bamboo glasses!

All of them, however, felt that working and living in bamboo huts was no great challenge; in fact, most felt that it was not only a novel but an enjoyable experience, not without opportunities for fresh forms of amusement! Placed in an uninhabited area, with dense jungle in the close vicinity and only a small village more than five miles away, their main recreation was shooting game in the patch of dense forest. One of Idris' colleagues was Reggie David, who had an excellent cook named Lal Khan whom he had brought with him from Kohat. Lal Khan had acquired some poultry consisting of a magnificent cock and two hens which were left to wander around, until he felt that the time was ripe for a special meal for his officers. One day, a young officer, Rozario by name, Idris recalled, who had gone for an early morning shoot into an adjoining patch of forest, came running back, holding up a bird and joyfully

announcing that he had shot a wild fowl. He soon had to be told that it was no wild bird, but one of poor Lal Khan's prize pet birds.

Flight Lieutenant Minoo Engineer, Idris' Flight Commander, was a very fine officer, Idris always told me. He was professionally demanding, but very warm as a person and with a great sense of humour. Idris, who loved to play pranks, found Minoo the right one on whom they loved to try them out, as Minoo took it all very sportingly and always responded with a hearty laugh! One day, a month or so after their arrival at Bawli North, Idris' colleague, Mac Mckenzie and Idris decided on a prank and, for it, invited Minoo to join them for a shoot, which Minoo readily accepted. After wandering around for a while, Idris went and hid behind a large bush. As planned, Mac then ran to Minoo calling out that a large animal had pounced on Idris and had dragged him away. Shocked and dismayed, Minoo swinging his shotgun around, started aiming at every bush in which he suspected an animal might be hiding. He kept telling Mac that he had been feeling so proud that there had not been a single casualty in their operations, and now this disaster had taken place! Meanwhile, Idris was still hiding behind the bush, but with Minoo's swinging gun, he was, understandably, petrified in case Minoo, seeing something largish moving in the bush, might fire, mistaking him for the animal! Finally, as Idris saw Minoo turning away, he dashed out calling, "Minoo, Minoo, it's me – don't shoot!" Minoo was such a sport that they could fool around with him like this, even though he was their senior officer, and, in fact, their Flight Commander. Throughout the years ahead, till Minoo retired as an Air Marshal, Idris' relationship with him remained one of an enriching and cherished friendship.

Once, when Asad and Asgar, our two boys, already well into their teens, were sitting with us and listening to Idris talk about the war, I distinctly remember Idris patiently explaining that the element of surprise, achieved primarily by flying out as low as safety considerations would permit, was of critical importance. With their Hurricanes, fitted with 20mm cannon and the ability to carry 2 x 500 lb High Explosive (HE) bombs, their standard attacks were pulling up steeply in the vicinity of the target to facilitate steep dive bombing attacks, and, thereafter, strafing with their cannons. This was always followed, however, by a quick getaway, again at very low levels, to avoid the often heavy anti-aircraft ground fire. Idris once recounted to me an incident when he was flying, as usual, as No. 2 to his Flight Commander Minoo, during their operational flying in Burma, which, unintentionally, of course, nearly ended in a disaster.

Minoo was apparently leading an attack on a Japanese camp and, keeping in mind the terrain and the nature of the target, had decided on a steep dive bombing attack with their HE bombs. In the four-aircraft formation, Idris was briefed to fly (as Minoo's No. 2) immediately behind him in the dive. The formation was briefed by Minoo that, as soon as he (as the leader) spotted the target, he would alert the formation and then first carry out a dummy dive in which all should follow and pull out, and then again go in for the final bombing attack. On sighting the target, Minoo winged over and went, as always, into the steep dive. Idris, as his No. 2, presuming that, as earlier briefed, it was a dummy dive (and bombs would not be released), followed close behind. To his horror, Minoo called out, "On target, attacking" and released his bombs! Idris, in his aircraft close behind, was suddenly thrown up by the unexpected blast of Minoo's two 500-lb bombs. Though shaken, Idris managed to control his aircraft and gained height, and rejoined the formation behind his leader.

Idris followed this by asking Minoo that, as it was to have been a dummy run and the rest of the formation had also 'peeled' off without dropping their bombs, could he, Idris, repeat the attack? Idris, of course, realised that Minoo too must have, by then, realised the error and felt terrible. However, in spite of the ensuing ack-ack ground fire from the now-alerted enemy, Idris repeated the dive and bombed the target. The other section of two Hurricanes was also able to drop their bombs. Happily, on returning to base, at the debriefing, a signal was awaiting them to commend the squadron on another very successful mission! Idris and the other two pilots of the second section, knowing how badly their much-loved Flight Commander would be feeling, later went to Minoo's office and Idris gave him a tight hug, saying, "All's well that ends well!"

Idris told me that, greatly to their delight, the squadron generally was given at least 2-3 such bombing tasks almost every day, each involving a four-aircraft mission, which meant plenty of operational flying for all pilots. Within a month or two, they had further honed their skills and gained confidence.

One morning, to his surprise — and delight — Minoo called Idris to his bamboo hut which was the Flight Commander's office, and told him that a fresh task had been assigned to the squadron, and he wanted him (Idris) to lead the four-aircraft formation as he felt that by now Idris had enough experience. He, however cautioned him that, as the target area was much deeper inside enemy-held territory, the Hurricanes would,

in all probability, have to refuel on their way back at a landing ground where just basic refuelling facilities would be made available. He should, therefore, prepare for it adequately and brief the others in his formation. Idris was, of course, thrilled by the confidence being reposed in him. As the long route, largely over enemy-held territory, was thickly forested and with very few landmarks, navigation at a low level posed an added problem.

For maximum surprise, as calculated and planned, when they pulled up for the attack, the target — a longish bridge — lay right ahead. The four aircraft now quickly, as briefed, changed to spread a little to ease correct positioning of each for the final dive in echelon starboard (right). In no more than a minute, all four aircraft, one by one, dived and dropped their HE bombs. “Red Four; bombs away!” called the last of the four aircraft. Fortunately, none of their aircraft was hit by the ensuing ack-ack firing at them. Idris glanced back to do a quick visual check of the accuracy of their attack and was delighted to see that the area of the bridge was obscured by smoke and dust and had, in all probability, been badly damaged and, hopefully, rendered unusable. Idris waggled his wings to call the other three aircraft back into their original two-section ‘finger four’ formation and set course, as briefed, for the designated airfield to refuel.

On arrival over the landing ground where they were to land and refuel, Idris noticed, to his dismay, that 4-5 cattle were grazing on the runway, rendering it unusable, at least temporarily. A quick glance at his fuel gauge convinced him that without refuelling there was very little chance of making it to their base at Bawli. Quickly trying to think of the best alternative, he called the other three aircraft to gain height and orbit whilst, as he told them, he would try a few very low passes over the runway in an attempt to scare away the cattle. To his relief, he was successful; and he also saw a few airmen in uniform with trucks on one side of the runway. Calling the rest of his formation, he told them that he was going in for a landing and, spacing themselves out, they should follow. Once all of them were on the ground and safely parked for refuelling, another surprise awaited them. There were no refuelling bowzers — there was more than enough fuel, they were told, but all in five-gallon ‘jerry cans’! However, with everyone lending a hand, in about an hour, all four aircraft had enough fuel to make base well before sundown! And, finally, came the icing on the cake: at the debrief after landing, a message was awaiting them, stating, “Bridge no longer usable — well done!”

Idris told me when I asked him about this time in Burma, the primary task of No. 3 Squadron was ground attack in support of our own and Allied ground forces engaged against the Japanese. Targets to be attacked included enemy troop concentrations, ammunition dumps and enemy lines of communication, especially bridges, in support of the Allied forces which were primarily our own formations. In the air, the Japanese had their own very lightly constructed but highly manoeuvrable aircraft, the "Zero", in substantial numbers. Idris said that in air combat, they completely outclassed our Hurricanes to the extent that strict instructions were given to his squadron that, if they sighted the Zeros, no attempt should be made to engage them in combat. This instruction, though, was not always followed, and on one occasion, a Zero was successfully engaged and shot down. Happily, instead of being reprimanded, the pilot was awarded the DFC! However, to minimise the chances of detection, the camouflage-painted Hurricanes always flew out and back at very low levels, mainly over the thickly forested area of land operations.

Apart from the unlikely event of being spotted and engaged by the Zeros, the main danger for the low-flying Hurricanes was from the heavy concentration of the enemy ack-ack guns at all critical points. An added danger was the terrain over which they always had to fly: which included hills and valleys, all thickly forested areas where, in an emergency, no forced landing would be possible. The drill, therefore, was that if an aircraft was badly damaged, or critically short of fuel, the pilot should pull up, gain a little height and then eject. Parachutes were, therefore, invariably equipped with survival kits consisting of some specially prepared items of food and a weapon. Special lectures on survival techniques were also a regular feature of their ground training, including information on the kind of wild herbs, plants and even insects and reptiles that could (or should not!) be eaten!

As Idris mentioned earlier to us also, the element of surprise, achieved primarily by flying out as low as safety considerations would permit, was of critical importance. A quick pull-up to a sufficient height of a few thousand feet that would permit steep dive bombing to help ensure maximum accuracy was always preferred. This was followed by a quick getaway to avoid the heavy ack-ack. Though some aircraft did occasionally return with hits, throughout their operations, the squadron did not lose a single aircraft. In fact, they were overjoyed when, during the final days of the war, the Commanding Officer , Squadron Leader Shivdev Singh, received a special

congratulatory signal from Lord Mountbatten, the overall Commander of the Allied Forces in that area. Flight Lieutenant Minoo Engineer, Idris' Flight Commander, was also awarded the much coveted DFC, much to the delight and satisfaction of the entire squadron.

In mid-1945, when the Japanese forces finally surrendered and the war was over, Idris, happily for him, was still with No. 3 Squadron, and under the command of Squadron Leader Shivdev Singh and Flight Lieutenant Minoo Engineer as Flight Commander. They flew back to India to St. Thomas Mount airfield near Chennai (then Madras) as their base. Weeks later, it was announced that VJ Day (Victory against Japan) would be officially celebrated in town on the following weekend with, to their delight, a holiday on Saturday as well. Their delight, sadly, was very short-lived. To their disappointment, disgust and understandable resentment, the squadron was informed that, to avoid the risk of their misbehaviour over that period in public places in town, all personnel of Indian units (including No.3 Squadron) would not be permitted to 'book out' and would remain confined to camp — a grim reminder, indeed, that India was still no more than a British colony!

From 3 Squadron, on 6th October 1945, Idris joined 9 Squadron, under the command of (then) Squadron Leader Asgar Khan (much later, after Partition, the Chief of the Pakistan Air Force) based at Risalpur in the Northwest Frontier Province. Fortunately for Idris, the squadron was equipped with Spitfire VIIIs, which he just loved to fly.

A year after the end of World War II, Idris was again selected to go to England as part of the Indian armed forces contingent to participate in the World War II Victory Parade to be held in London in June 1946. This time, he was one of only three officers and 21 men who were picked for this. A very special event was the visit to their tented camp in one of London's parks by the Queen, accompanied by her two daughters, Princess Elizabeth (presently Queen Elizabeth) and Princess Margaret, to whom, especially, Idris was introduced. After the parade, they were flown to Berlin and Cologne to see, first-hand, the extensive destruction caused by the bombing raids of the Allied Air Forces during the war.

Pilot Officer, Idris Latif being introduced to Princess Royal (sister of King George VI) after Victory Parade in London, June 1946 (Sqadron Leader R.H.D. Singh, leader of the team on the right)

L to R: Princess Elizabeth (now Her Majesty Queen Elizabeth, Princess Margaret, Queen Elizabeth talking with Idris

OCCUPATION FORCE IN BURMA

Spitfire being flown by Flying Officer Idris Latif, Burma, 1946

On their return in January 1946, Idris rejoined 9 Squadron at Risalpur, still under the command of Squadron Leader Asgar Khan. To his delight, Flight Lieutenant Akhtar Khan, of whom Idris always speaks in very warm terms, was his Flight Commander. Within months, the squadron again flew to Burma, but this time as part of the Allied Occupation Forces.

On the way, Idris had a rather painful experience. En route, the squadron had landed at Cox's Bazar airfield for refuelling, and whilst this was in progress, all the pilots' decided to go into the canteen for a quick cup of tea and a biscuit. As Idris, with the rest of the pilots, was walking in, he saw that one of the Royal Air Force Sergeants sitting in the canteen had a pet monkey on his shoulder. Idris decided to keep a good distance as he walked by. But just as Idris passed, for no reason, the monkey decided to jump onto his shoulder and, worse, much worse, bite his arm. Of course, the monkey was quickly taken away, but not before it left a nasty and somewhat bloody mark on Idris' right arm. A doctor was immediately called and, after dressing Idris' wound, told him that with a monkey bite, there was a serious risk of rabies and it would be best if Idris stayed back for a few days with them for treatment and observation. With the whole squadron preparing to take off, Idris refused. The doctor, seeing how determined Idris was, said that he (Idris) would be doing so at his own risk, warning him also that if on the flight he found himself frothing at the mouth, he should land — if necessary, force-land — without delay. Idris insisted on

flying on with the squadron, and happily, without any problems in flight. On landing at their destination, the squadron doctor took charge and insisted on a daily injection for no less than 14 days. What was worse, Idris recalled, it was a painful experience, especially as each was no less than a 5 cc injection given in his stomach, which, even though well massaged, invariably remained painful — particularly, as Idris smilingly added, whenever he laughed! Idris, therefore, requested his squadron friends, "as a very special favour to me, no jokes in my presence for 14 days, please".

This time in Burma, the squadron was based at Mawbi — "I think, that is how it was spelt," recalls Idris — an airfield close to Rangoon. Only tented accommodation was made available — something which, however, by now, the Air Force had become well used to. When the weather turned warm, all of them preferred to sleep in the open outside their tents on cots but, because of the very many mosquitoes, always, each had a personal mosquito net. One evening there was a very heavy downpour; even after the rains stopped, the ground remained completely waterlogged. Soon after, when they all got into their beds with the mosquito nets carefully tucked in, to their horror they saw big dark scorpions crawling up the nets on all sides. Apparently, they were later told, the scorpions were trying to escape from the water all around. Idris, like the others, moved into the centre of his bed, taking care to keep well away from the net. Sleep was, of course, out of the question! With the first light of dawn, to their great relief, they saw that, as the water had drained off, the scorpions had also gone! Happily, none of them had been stung; but, it had been, understandably for all of them, Idris said, a rather terrifying and, certainly, an unforgettable experience.

As mentioned earlier, Idris' Commanding Officer was Squadron Leader Asgar Khan and Akhtar was still his Flight Commander. In the squadron also was Nur Khan who, Idris said, was perhaps not only the most skilled but also the most daredevil pilot he has come across. More than once, Idris said, he saw him, when coming in to land, turn the Spitfire upside down, and then select the undercarriage to go down (in this case, of course, it would go up!). Still on his back, he would turn for the final approach, quickly selecting flaps and landing — and always it was what we often called a 'kisser' i.e., a very smooth and perfect landing — just breathtaking to watch, according to Idris. Idris' own record in flying, incidentally, as I find in his log book was, throughout his career, "Above the Average"; and even marked "Excellent".

Generally, at that time, each pilot had an aircraft allotted to him to look after, and if serviceable, fly. All went well, except for one very shocking error on Idris' part. One day when he had been detailed to fly, he was told that his Spitfire was unserviceable. Akhtar, without hesitation, said "Take mine", adding "but take very special care". Idris had flown it once before and knew that it was truly a wonderful aircraft to handle. Once airborne, he had a great time doing aerobatics with it, thinking to himself that it was such a joy to handle, that on landing, he would tell Akhtar that it was really one of the finest Spitfires he had ever flown. Knowing also that Akhtar may well be watching him, Idris decided that when he came in to land, he would, 'show off' (as many, if not all, particularly fighter pilots, unfortunately, love to do), by doing an unusually tight circuit before landing.

Keeping close to the airfield and turning in from downwind as tightly as he could, Idris, whilst also putting down the undercarriage, straightened off for the finals just in time for the flaps to be operated before touch down. Full of *joie de vivre* (!), all went perfectly, except, tragically, as they found later, due to a technical malfunction, the undercarriage didn't lock fully! To his horror, he soon, very soon found, as he touched down before coming to a stop, there was a horrible screeching-grating sound under him, with the belly of the aircraft and the collapsing undercarriage very badly damaged. The propeller was also, of course, shattered, and there must have been substantial damage, he felt certain, to the engine and the ancillaries. "God, what am I going to say to Akhtar" was all that Idris could think about, as he sat quietly, shocked and, of course, deeply distressed by the catastrophic end to a wonderful flight — in a truly wonderful Spitfire. But, Akhtar, being the wonderful kind of man that he was, and, of course, knowing what Idris must be going through, only said, "Too sad, Idris, I know what you must be feeling. Don't worry, it can happen to any of us; but God has been kind, and I'm glad you are not hurt." What a man, Idris thought to himself. No wonder, Idris was fond of him.

Idris often mentioned that not only when in a squadron but relatively junior officers are often allotted, in addition to their regular work, 'secondary duties', when posted at squadrons or stations. A challenge for Idris was when Asgar Khan, being his Commanding Officer, gave him the 'secondary duty' of teaching the Lascars Urdu — something he never thought he would ever be asked to do. Fortunately, seeing the keenness and enthusiasm of the Lascars, it turned out to be, Idris said, a fairly emotionally rewarding experience, adding with a smile, "How much they learnt is another matter."

CHAPTER

07 Selection Board, Dehradun, During Partition

Soon after their return to India, Idris was posted to the Air Force Selection Board in Dehra Dun, one of two or three centres where those wanting to join the Air Force as officers in the Flying Branch went through tests to assess their suitability. The applicants, usually in batches of about 25 or so, would be divided into 'syndicates' of around seven, and, during the three days that they were there, would be given a series of tests. These included one or two individual tests, such as giving a brief talk on a subject presented to them just a few minutes before. More importantly, there were a few 'group' tests consisting of getting through, but essentially as a group, a series of obstacles, of which two or three were quite difficult. During these, all the youngsters were under the exceptionally close observation of those whose task it was to assess them. Almost invariably, this was the President of the Selection Board, a GTO (Group Testing Officer) of Flight Lieutenant rank (Idris being one of three or four others) for each of the other syndicates. In addition, there was always a highly qualified psychologist/psychiatrist.

The primary task of this small assessing team was, in a nutshell, to individually observe and note the personal qualities/character of the applicant. The main emphasis was not so much on the applicant's ability to think out and go through the obstacles as quickly as possible as an individual, but on his ability to think out ways and lead the whole group through the challenge as quickly as possible. In other words, it assessed quick and logical thinking, cooperation and team spirit and, of course, of particular importance, his capacity to lead. Each officer on the testing team was required to keep

his observations and analysis strictly to himself until the final conference on the third day. Under the Chairmanship of the President, each, including the President, would then present his personal assessment. What was very remarkable, Idris once told me, was that he found there was surprising commonality in the assessment of each officer — something which seemed to underline the validity of the tests and the accuracy of the assessments.

Idris often spoke about this as an invaluable education for him, a great lesson in how to assess the personality, character and instinctive reactions of individuals, particularly when under stress. This experience, Idris said, was invaluable throughout his later years in the Air Force, since a good and reliable assessment of the capabilities and limitations of those serving under one is obviously of critical importance. His boss and his colleagues were all also wonderful people and his time with the Selection Board has left him with very happy memories, except that it was during that period that the country was partitioned, and with it, the armed forces.

Very soon after Partition was announced, Idris, whilst still at the Selection Board in Dehra Dun, was most surprised to receive a call from an officer who was leaving for Pakistan. He spoke to him whilst he was still in Delhi, expressing the hope that Idris, being a Muslim, would also move with them to Pakistan, adding as an incentive, Idris thought, "In Pakistan, we will be small in numbers and, therefore, promotions will be very fast!" Idris laughed and said, "You must be joking; what has religion got to do with nationality? I'm an Indian, born in Hyderabad, you know — and not the Hyderabad in Pakistan! But, don't tell me that you are leaving?" Idris laughingly added, "With your skills and well-deserved reputation, you could well be on an even faster track with us here, in the Indian Air Force! In any case, I'm sure you'll agree, in life, career is important, but one's country is one's country, and I am, not just happy, but more than happy to remain where I am." After narrating this to me, Idris finally added, "That was more than half a century ago. Since then, we have had no contact — ever. If he is happy, I am happy; whilst India was one, he was a valued colleague".

Partition resulted, as is well known, not only in much loss of life, but also terrible suffering even for those who survived. One consequence was that, of the many who migrated to India, quite a few came for security to Dehra Dun; but,

happily, everything remained generally peaceful. One morning, to Idris' great surprise, Air Marshal Subroto Mukerjee, then everyone's much loved and respected Chief of our Air Force, personally telephoned Idris to ask if he would prefer to be transferred to a safer place. He felt that being a Muslim, Idris' life might be at risk because of the reported disturbances, adding, "Idris, these are difficult days and there are, unfortunately quite serious disturbances in several parts of our country; I hear that Dehra Dun has also become somewhat unsafe. We are, of course, most happy that you have decided to stay on with us — and quite honestly, knowing you and your family, we are not surprised. But I want to make absolutely certain that nothing happens to you that may leave a scar. Please feel free to tell me where you would prefer to be — at which station, I mean — during this tragic period, and I will arrange it." Deeply, very deeply moved by his Chief's concern — and his Chief's most caring gesture of personally speaking to him — Idris said that he was deeply touched by this very special and most thoughtful gesture; but, he added, he was absolutely safe and with wonderful comrades and friends, and that he would be happiest if he was allowed to complete his time at the Selection Board. Idris was delighted that his request was acceded to.

There were a few incidents off and on; not, of course, on the camp but in town. For instance, one morning, Idris and his very close friend, Kalyaniwala (or 'Kally' as he was affectionately called), spotted some ladies riding cycles getting down in front of the bigger stores, going in and loading up selected items on their bicycles. With (unloaded!) pistols, but doing their best to look aggressive, Idris and Kally always managed to frighten the looters away! However, although there was absolutely no danger to any defence personnel, including those at the Selection Board or at the Military Academy in the vicinity, there certainly were disturbed conditions. The road between Clement Town, where they stayed at the Air Force Selection Board camp, and Dehra Dun, became so empty in the evenings that, after working hours, young officers got opportunities to indulge in really fast driving and racing their motorcycles. Cars were, of course, out of their reach till many, many years later!

On completing his term at the Selection Board, Idris was posted to Barrackpore and given command of 'A' flight of No. 10 Squadron, which, to Idris' delight, was also equipped with Spitfire 8s, which Idris, as I have often mentioned, loved to fly.

Strangely, for some reason, both Squadron Leader Rehman, the then Squadron Commander, and his successor, spent little time in Barrackpore. Idris didn't mind this at all, he said. In fact, he was, he confessed, quite delighted, because, in their absence, as the senior-most officer, he had a wonderful opportunity to be in overall charge, not only of the squadron, but of the whole Barrackpore Station, especially as he had the freedom and authority to fly as much as he wanted!

CHAPTER

08 Playing with Fire

In 1948, Idris was posted as the Chief Ground Instructor at the Initial Training Wing (ITW) Coimbatore, then the initial ground training establishment for all pilots of the Air Force. With a fine Station Commander (Wing Commander Sen), an excellent staff, and a whole lot of young, enthusiastic wonderful cadets, life was stimulating and emotionally rewarding — except that, being a Ground Training Institution, there were no aircraft to fly! Two events stand out in his memory of that period. Since there were no aircraft, Idris would go to Bangalore as often as he could, for a few hours of flying on Harvard aircraft with the Communication Flight of Headquarters, Training Command. On one occasion, he decided, he told me years later, that, as all the trainees at ITW were potential pilots but had never seen even a trainer aircraft of the Air Force, it would be a tremendous motivation for them if he flew to Coimbatore and did some aerobatics right over ITW. This he proceeded to do on one occasion, and then he was flying back after the aerobatics and passed over the civil airfield at Coimbatore. As he flew over, he spotted a civil airlines Dakota which had just landed, and from which the passengers were deplaning. As he confided to me much later, "I just couldn't resist making a really low pass over them" and was thrilled beyond expectation to see some of the passengers running to seek shelter under the wings of the Dakota. Greatly tempted to make just one more pass, he was turning around when, fortunately, Idris said, his sense of discipline triumphed over temptation and he just flew past waggling his wings to convey, in Air Force style, his greetings and good wishes.

Landing soon thereafter, he quickly got into his jeep and drove back to the camp. He had hardly got to his office, when not unexpectedly, Wing Commander Sen, his

Commanding Officer, whom he not only respected, but with whom he had a warm relationship, called him to his office. "Idris, what have you been up to? I've just had a report from Colonel Lesley Sawhney, the Deputy Commandant of the Staff College in Wellington, that an Air Force aircraft was beating up the airfield, doing

Pandit Jawaharlal Nehru addressing the cadets at the Initial Training Wing at Coimbatore, 1948

In the mornings, I knew that he flew sorties in his Tempests almost every day. I would often sit at a radio in our room and keep twiddling the controls till I latched on to the garbled sounds of the call signs of Idris and of the other fighter pilots who were flying. Their words were not clear but just the sound somehow reassured me, for they were not calling out “May Day’ which was, Idris had once told me, the call when things went wrong! After the air crash of Idris’ best friend Ken, I was really terrified of Idris’ daily flying. One afternoon, he got home later than usual. By that time, I was out on the main road, pacing up and down anxiously, until he arrived in his official Jeep. When he saw me looking so anxious, he told me with great seriousness that if he felt that I was worrying so much, it would remain very much in his thoughts when flying, whereas he should be concentrating solely on his flying. I, therefore, decided from that day to pull myself together and switch off that side of me from then onwards. I don’t think most people realise how Air Force pilots, especially fighter pilots, can never be sure of returning home after the daily manoeuvres they perform, sometimes even at supersonic speed, while practising their offensive and defensive manoeuvres and other complex, often risky exercises. I got to know later that many of the wives developed serious headaches through their constant nervousness and worrying.

Ken’s tragic end and the thought of his widow Iris continued to haunt me. Nothing could replace her beloved husband, and she had no home or a pension that would suffice for even her daily expenses. For decades, this was the plight of the widows and children of the Air Force officers who died in the service of the nation. Very few girls in those days were educated to take up a career. They usually got married after graduating from college or sometimes even earlier, and many widows were not qualified to earn enough to even afford to educate their children in good schools. Officers usually did get themselves insured, but would sometimes neglect to nominate their wives, in case of death, as their beneficiaries. The nominees would often remain a parent. At times, both were financially needy. It was a matter which we took up seriously after the Air Force Wives Welfare Association (AFWWA) was formed more than two decades later. But, as in the case of Idris’ friend Ken David, those at the Air Force stations where the officer had been posted always rallied round and did what they could to help the widows. In subsequent years, during our married life together, I learnt what a deeply moving and comforting experience such caring could be, but nothing could replace the loss of the loved one.

From Pune in April 1951, Idris was selected and assigned to go with Air Marshal Ivelaw Chapman, the Air Chief, on a goodwill mission to Thailand and Singapore. Very soon thereafter, when he returned, he learnt that with effect from June 1951, he had been posted to the Joint Services Wing at Dehra Dun as the Squadron Commander of C Squadron. Idris was, understandably shattered to learn that his command of a Fighter Squadron would very soon be over. But as Air Marshal Ivelaw Chapman had mentioned to me when he was on an official visit to Pune, "I know that every fighter pilot's dream is to command a Fighter Squadron and your husband has done well. But I assure you that he will find his next assignment very rewarding too. It is also a good, peaceful assignment for a newly married flyer!"

Joint Services Wing

Although Idris greatly missed his regular daily flying in his squadron in Pune, it was for us our very first home, and we had a very happy time at the Joint Services Wing (JSW). The JSW was located at Dehra Dun in the north, where winters, we soon found, were pretty cold — though enjoyable also. Incidentally, some years later, the JSW became the National Defence Academy of the Armed Forces and was relocated at Pune. At the JSW, General Wadalia, a fine, warmhearted officer and a gentleman to his fingertips, was the Commandant with overall responsibility of the Army's renowned Military Academy as well as the JSW. The latter was the direct responsibility of Captain Reggie Sawhney of the Navy, as its Deputy Commandant, another charming and friendly person with a wonderful sense of humour. As an example, which Idris recounted to me, was an occasion at a very formal Annual Day sports function, when the Chief Ground Instructor appeared on the grounds, immaculately dressed in a suit with even a red rose pinned on the lapel of his coat. On seeing him, Captain Sawhney quite loudly remarked to the delight of all within hearing, "Ah! There comes our Major, heavily disguised as a gentleman."

Under Captain Sawhney was the Chief Instructor, (the late) Wing Commander Rammuny, who was responsible for four squadrons: A, B, C, and D. Idris was in command of C or 'Charlie' Squadron. After two successive years of commanding C Squadron, when, among the squadrons, it held the championship over both years, Idris' suggestion of renaming it "Champion" instead of "Charlie" was happily accepted at all levels. Thenceforth, 'Charlie' Squadron became known as 'Champion' Squadron!

Major (later to become General) Krishna Rao was then commanding D Squadron. All Idris' seniors and colleagues were, I found, warm and caring and, very soon, made me feel very welcome. In fact, General Wadalia would refer to me as the "Honey from Hyderabad". I got fully drawn into squadron activities, sprucing up the Cadets Mess for the cadets, who, incidentally, were almost my own age, and soon they felt completely at ease — as did I. The wife of the Commandant of the Wing suddenly informed me one day at a Ladies Meeting that I was to be in charge of the Arts and Crafts stall at a forthcoming Fair. I was totally flabbergasted and I still cannot recall how I was finally able to stock up the stall and make it a success. Of course, I suppose it also did much for me to gain confidence in shouldering, so to speak, semi-Service assignments, but it was an awful ordeal to have to undertake this at the time.

For our living accommodation, we were allotted a barrack of five rooms in a row, which were more than enough. These, as all other buildings on the camp, we soon learnt, were earlier used as barracks for the Italian prisoners of war during World War II. Apart from the five rooms in a row, with a verandah in front and a kitchen plus a staff quarter at the back, we had more than enough space to spread out. Shanker, Idris' valet at Pune for several years, had fortunately agreed to come with us. Luckily, we soon got a cook who turned out to be an expert, I found to our delight, with knowledge of a variety of rather sophisticated Western dishes also. The only major drawback of our accommodation here was that there was no running water and only dry sanitation — quite a shock/surprise and a change from the comforts that Idris and I had taken for granted from the time we were born! But, happily, we both soon got used to it, feeling embarrassed only when guests came to stay with us — the first being Idris' parents. Fortunately, they also very sportingly took it in their stride. I remember that we bought some inexpensive material and I sat and made curtains, and also fixed covers for the sofa and chairs which were in the house. I blessed my stitching classes in school where we had been taught basic sewing and embroidery, and had to pass an examination by the famous Good Housekeeping along with the Junior Cambridge Exam from the UK.

We had an airy and comfortable home with a small garden patch in front, where our flowers grew in abundance as the soil and climate were both conducive. I especially recall the cosmos which grew lush and taller than my five-foot eight-inch height! Antirrhinums and nasturtiums in graded beds of lovely flowers were soon

blooming around a green lawn with a lovely tall and graceful eucalyptus tree to one side. Idris, being a very good as well as a keen rider, was also able to ride often. As there were good horses there, some especially for the initial training of the cadets, I also got initiated and acquired some expertise which I was able to use later in our Air Force life. Till then, I was quite scared of horses as I had fallen off a horse when I was a child, even though the horse was stationary!

The areas around Dehra Dun were lovely for drives and especially for picnics. The road to Mussouri, which we occasionally visited, went winding through lush, green forested hills, though the steep turns were scary to drive on. As we drove higher, the aroma of the tall pine trees pervaded the air. There is something so breathtaking about the view of the mountains and of looking down into the valley far below. One could see the rivers with flowing water and, on the banks, animals grazing on grassy slopes near little houses with their red-tiled rooftops. The road to Delhi was also along forests, especially beautiful after the rains, and later when the foliage of the trees got covered with the rich bronze and rust colours of their autumn leaves. On a holiday, we would pack a picnic breakfast and drive off to find a picnic spot, sometimes along a stream with its water shimmering in the early morning sunlight. We were able to take several guests along, including Idris' sister Surrayya, her husband, Badr Tyabji and their children, Idris' parents, his uncle and aunt, the Hamid Alis from their beautiful home in Mussouri and Begum Hamid Ali's brother, Salahuddin Tyabji, and his wife, Akthar, from Hyderabad. Hamid Ali *Chacha* had retired from the Indian Civil Service (ICS), but Begum Sharifa Hamid Ali was an active President of the All India Women's Conference. She and her brother were the daughter and son of the freedom fighter, Abbas Tyabji. They were all very intellectually oriented and up-to-date on reading and, each, we soon found, had a great sense of humour!

CHAPTER 11
To Delhi as Deputy Director Weapons

In July 1952, Idris went on promotion as a Wing Commander to Air Headquarters in Delhi as Deputy Director (DD) Weapons and, soon thereafter, was asked to take over as DD Operations. It was remarkable, we have often thought, how strong is the secular fabric of our society and our government that even when our relations with Pakistan were still far from 'neighbourly', a Muslim could be put in charge of top secret operational plans and documents. Idris enjoyed the trust of not only the top levels of our government but also, within the Defence Services, from the Air Chief, Air Marshal Mukerjee, downwards. Very much later, also, at the time of the 1971 conflict, I remember that the Air Force Chief (then Air Chief Marshal P.C. Lal) asked Idris to lead a small team to visit all the frontline bases and be his point of contact in any emergency when time was critical. In fact, I remember in the final stage of that war, when the Air Force launched a devastating attack at East Pakistan's operational centre that forced the Pakistani leadership to surrender, it was through Idris that the Air Chief routed his operational directions for Eastern Air Command. This was a reflection of the country's and Idris' secular standing. Even at the highest level, though a Muslim, all deemed Idris not only fully competent but also completely trustworthy in the planning, management and control of even the most critical operational roles and tasks that were not infrequently assigned to him, first as DD Weapons and immediately thereafter as DD Operations.

It was in 1952, when Idris was still DD Weapons that he took the decision that the rather unique but quite deadly napalm bomb, extensively used during the latter stages of World War II, should be developed on high priority in India for our Air

Force. However, Idris recognised that because of its rather inhuman dimensions, there could be a lot of resistance to this decision at a political level. To this, his argument was that war itself is an inhuman activity; to be killed or maimed by a bullet or a conventional bomb would be no better. Unfortunately, though the technology was simple and straightforward, the Defence Research and Development Organisation, to which the task was entrusted, wanted to import a few of these bombs so that they could, euphemistically phrased, adopt 'reverse engineering'! Since Idris felt that it was a matter of some urgency, he arranged for the import of a few. To test the political acceptance of such a weapon, Idris had had it included as an item in the firepower demonstration that had been arranged at the Tilpat range in the vicinity of New Delhi, and to which the Prime Minister, Pandit Jawaharlal Nehru and other very senior dignitaries of government, were invited. Air Marshal Aspy Engineer was then the Deputy Chief of the Air Staff, and Air Marshal Subroto Mukerjee was the Air Chief. At the time, Idris, of course, was just a Wing Commander. But clearly even Wing Commanders had considerable standing (!), as his suggestion was promptly accepted at every level, and the napalm bomb became, for that time, a valued item on the inventory of the weaponry of our Air Force.

When Idris moved to Delhi, I had to stay on in Dehra Dun till accommodation was made available in Delhi, while Idris stayed at the Central Vista Officers' Mess. One night a thief tried to break into the bedroom, and I screamed. Our brave Garhwali domestic employee was sleeping in the drawing room and when I yelled out to him, he dashed outside to catch the man, who ran away and escaped. Within a few months, I was delighted when we were allotted a flat at the Central Vista Mess for officers' families. In Delhi, after setting up the apartment which became lovely with our carpets and *objets d'art*, it was soon time to go to my parents in Hyderabad for the birth of the baby. I had become anaemic and the doctor, whom my parents arranged to examine me, said that I needed to have an intensive course of calcium and Vitamin B injections. With all that loving care, I was perfectly well when, to our great joy soon after, our lovely daughter, Mariam, was born in Hyderabad on 14th October 1952. She seemed to all of us to be the most beautiful child that we had seen.

In the three years that we remained in Delhi, we again changed residence twice, and then, finally to a spacious house at 42, Lodi Estate. It had a lovely big garden with a vast grassy lawn all around, where we soon grew large beds of flowers. Another great

joy for us was when our darling son Asad was born on 25th February 1955, and was a happy, healthy and smiling little boy. Happy, very hard working, and smiling he still is, though no longer a little boy.

Often on Sundays, we went on picnics combined with partridge shoots beyond some of the nearby villages. Duck shooting was less frequent and entailed driving out before dawn at about 3 am in the bitter cold of winter. When we would arrive at a lake where there were lots of duck, Idris would actually wade into the icy water to wait for the duck to fly overhead, and then he would fire. Our little Mariam and later Asad too would be all wrapped up, and the three of us would be sitting in the car till daybreak when the warming sun came out, but Idris would be standing in freezing cold water. I remember that the first time we went on a partridge hunt, and the birds flew out and were fired at and shot, tears used to roll down my cheeks. Although I felt bad about shooting birds or animals, they were never unwelcome for the pot! Idris, I have to say, was very, very particular about going for our shoot-cum-picnics strictly only during the 'open season' for shooting, when it was officially permitted.

CHAPTER

12 Across the Seas, A Hand of Friendship

Idris was sent to Indonesia in December 1955 as part of a three-member team to advise and assist the Indonesian Air Force (or AURI, as it was known) to induct their first jet fighters. Indonesia's relations with Pandit Nehru were, of course, well known, and they had turned to the Indian Air Force, we were told, as a little earlier our Air Force had very successfully inducted our first jet fighter, the Vampire.

Ambassador Tyabji extreme left, Mrs Tyabji third from right, President Sukarno, and Bilkees Latif, extreme right

I was also delighted when we had been told that, since the assignment would be for at least a year, our two children, Mariam, then five, and Asad, not even two and I, would be accompanying Idris. An added boon was that Surayya Apa (one of Idris' elder sisters) who was married to their cousin Badr (Tyabji) bhai, who was then our Ambassador at Jakarta, and we looked forward to seeing them there.

En route, a big thrill was when Asad, who was ten months old, took his first steps on the deck of the ship before we reached Indonesia. After we got to Jakarta, Mariam joined a school near the house. Although she went there quite happily with me, when the teacher insisted that I should not stay, she wept and wept. I could hardly bear to leave, and as I walked away I could hear her sobbing almost all the way back.

Besides Idris, the other members of the advisory group for the AURI consisted of the leader of their three-member team Group Captain Anantha Narayan, who was a bachelor, and Wing Commander Khanna, with his wife, Usha, their son and their baby daughter. To accommodate all of us, we were allotted a large house with a side wing, a garden and staff quarters on Djalan Budikemuljan. It was not far from the President's palace and the large open grounds of the Merdeka Medan. Our stay in Indonesia was a unique experience of warmth extended by everyone there, from President Sukarno, the Air Chief Suryadarma and his lovely wife, the former Prime Minister Shahrir and others. The very fine intellectual Sujatmoko, who was later the Indonesian Ambassador to the US, as well as Air Commodore Abu Rahman of the AURI and several others, became special personal friends. Idris was the only member of the team who was from the Flying Branch and with personal experience on jet fighters. He was, therefore, kept very busy, but happily so, in the task of formulating plans and programmes for the basic training and operational employment of their very first batch of jet aircraft. This was the Vampire — a type that our own Air Force had acquired a few years earlier, but on which Idris had acquired considerable first-hand experience. It was, therefore, both a challenging and satisfying task, made particularly so as the AURI authorities and officers were deeply appreciative of his work — all of them, from their Chief downwards.

As I was soon able to converse in some Bahasa Indonesia, I used to go several times a week to help out at the Indonesian Red Cross Centre. I was also made Honorary Secretary of the International Women's Club. Often, at our parties, I dressed the lady

guests in sarees which they thoroughly enjoyed, while I enjoyed wearing the lovely Indonesian outfit with its *kain* and *kabaya*. Since we had taken our Vauxhall car with us, a special treat was to travel to a few of the other cities of this beautiful country, including Bandung in the mountains nearby. I remember that it was so lovely in this mountain resort that, at the AURI guest house where we were staying on a two-day visit, Idris, with his prized movie camera, decided to photograph some portulaca flowers as they were opening up in the sunlight. He fixed his movie camera on a stand facing a plant that was laden with flowers about to bloom. We knew that these flowers open up every morning in the sunlight. So, as the sun rose, and the flowers were slowly opening to fully bloom, Idris would patiently, every few minutes, press the exposure button (a technique that he had evolved). He also similarly filmed their closing at sundown. It required infinite patience, but was successful — the whole process of the flowers opening and then slowly closing for all of us to see, smoothly recorded on his camera. It was really delightfully spectacular to view it. Many years later, using the same technique, Idris again made similar movies of the flowers opening up into full bloom and closing, when we had our own lovely garden at Air House in Delhi, when he was the Air Chief.

The beautiful classical Wayang dance form was performed at special functions continuing almost through the night. We were amazed to see the great detail in which they depicted episodes from the *Ramayana* and the *Mahabharata*. Indonesia was a perfect example of the beautiful cultural bonding in Southeast Asia. Large numbers of people had Sanskrit names and followed Islam as their faith. President Sukarno, who knew several languages well, which included Sanskrit, explained to me once the meanings of his name and those of his family which were of such a combination. Our visits to Jogjakarta and to Borabudur and Prambanan were so enriching with their fabulous Buddhist and Hindu temples.

During our stay in Indonesia, thanks to the courtesy and hospitality of the AURI, Idris was also able to briefly visit the beautiful and hospitable Flores Island. On his return, I clearly still recall his telling me how very deeply moved he was by the bonds of understanding and exceptional closeness of the different communities — the Buddhists, the Muslims and the Christians. Particularly admirable was the unwritten understanding — indeed, claim — that should a temple, church or mosque ever need repairs, the other two communities claimed it as their right to repair it. It is so sad that,

today, in so many parts of the world, religion, instead of creating bonds of harmony, friendship and closeness, often — too often — has become a source of friction and rivalry.

Idris' work with the AURI was both emotionally rewarding and professionally stimulating. He was also delighted when the AURI Chief arranged that whenever Idris requested it, a Harvard aircraft should be made available to him for his flying practice. Only once was he faced with a crisis. Indonesia, as those who have visited that country know, has, amongst many other attractions, some particularly beautiful coastlines. As his hobby was photography, Idris usually had his camera with him. Just on one occasion, impulsively, he decided to take it up with him when he was flying, and if possible bring back some pictures to show me! After taking a few pictures, whilst returning to base (Jakarta Air Base), he smelt something burning in the cockpit. Soon, quite a bit of smoke started coming in and he realised that there was a fire somewhere in the aircraft, and with it, of course, a rapidly developing risk of a serious accident. His first action was to quickly throttle back. He was greatly relieved to find the smoke

Idris on the left, with Air Marshal Suryadarma, the Chief of the AURI, on the right

quickly lessening; but with it, he was also, of course, losing height. Since he was no more than just a thousand feet or so above ground, he had no option but to try and restore normal power. But this, he found, only increased the fumes in the cockpit. To make it worse, at that time, he was flying over an area covered with fairly dense forest which he knew was under the control of the opposition Communist Party. The thought occurred to him that if he crashed, and was found with a camera, particularly with photographs of the coastline, the consequences could be serious for him. Worse, if he survived, he could not even speak the local language to explain anything. He realised then, that before bailing out or crashing, he first had to get rid of his precious camera!

It was a beautiful camera, and photography had been Idris' passion ever since his childhood; so it was with a heavy heart that he put his hand into the pocket of his flying overall to take it out. Ultimately, he just could not bring himself to chuck out something so precious to him, and back it went into his pocket! All this time, he was heading back towards the airfield at Jakarta. He was getting lower, but managed to maintain sufficient altitude, by alternately throttling back to reduce the smoke; and again, when getting too low, restoring power to gain height. At the same time, unfortunately, the smoke in the cockpit kept increasing, and with it, he was fully aware, the risk of the aircraft going up in flames was increasing too! At one stage, he suddenly, to his delight, spotted an open field in the forest, and toyed with the idea of a forced landing. Coming closer though, he saw enormous goalposts on either end of the field which, he decided, made the approach for a forced landing too risky. The only option left then was to continue climbing, and then gliding, by the judicious use of the throttle, to reduce the fire hazard. He kept hoping, in his heart, that he could, thus, cover the remaining forty miles to the Jakarta airfield! Fortunately, it worked, and reaching Jakarta, Idris safely landed at the airfield. His precious camera also remained safely with him — for many, many more years. I was also able to see from his pictures a few more of the beautiful sights of that green and lovely country!

Months later, in 1957, whilst still in Indonesia, we got news that Idris' dear father had, very suddenly and unexpectedly, passed away after a severe heart attack. Fortunately, most of Idris' brothers and sisters were able to be in Hyderabad to support Idris' wonderful mother. It was tragic for all of them, but especially for Idris, who was so far away and also longed to be with her. He and both his parents

were very close to each other, and Idris' father had always been a very inspiring head of their family. His mother, most courageously, decided that she would not move in with any of her children, but would stay on in her home and visit each child and family in turn. She told them all that they should also visit her regularly, and to come as often as possible to stay with her at her home and give her the happiness of welcoming them.

ADVENTURE EN ROUTE TO STAFF COLLEGE, WELLINGTON, INDIA

On our return from Indonesia, we each spent a brief vacation with our respective mothers in Hyderabad. Idris then went by road, driving his car to Coonoor to join the Staff College course at Wellington. He had with him a person who was to work with us as our cook. The drive was unexpectedly fraught with some risk, as he soon discovered on the way, from which he was, luckily, saved by a good Samaritan's kindness and consideration. They reached Chennai safely, but later in the afternoon, on the way from Chennai to Coimbatore, driving with a lot of baggage in and on top of the car, after about 45 minutes, he stopped to check if the tyres and the suspension were in order. To his dismay, when he checked the rear suspension, he found that one of the elliptical metal strips of the suspension had cracked and was resting on the lower one. It was clearly a matter of concern and quite hazardous as they were making the journey on a relatively rough road, especially at night, with the extra weight of a large amount of baggage.

Idris, therefore, stopped at the next habitation to ask if there was any workshop in the vicinity where he could get the car repaired. A man immediately came forward and said that there was a place a little behind the village and that he would be happy to guide him there. Idris was just getting ready to follow him when a truck which was passing by stopped, and the driver, getting down, asked Idris if anything was wrong. Idris showed him the spring and said that he had been told that there was a small workshop behind the village, and that one of the villagers had offered to take him to it. The truck driver said that he had a better solution to offer. He added that there was a very good garage a little further, the owner of which he knew well; to enable Idris to get there, he would be happy to do a temporary repair on his car. Idris thanked him profusely but became a little sceptical when the truck driver produced a strong rope and tied the full length of the spring. He then asked Idris to drive on but a little slowly, at not more than 20 miles an hour. When Idris expressed his reservations, the driver

reassured him by saying that, to help in case of any mishap, he would be following in the truck, driving slowly just behind him.

Reassured, Idris, with the truck behind, set off. As requested by the truck driver, he stopped after about 15 miles so that the truck driver could check if his temporary repairs were holding. Finding it very satisfactory, but still as a precaution with the truck driver following behind him, they drove on. Finally, after some hours of very careful driving, with the truck then leading the way for the final five miles or so, at about 2 o'clock at night, they reached the garage. The truck driver went in, woke up his friend there, and told him that Idris was a military officer who had to reach Coimbatore by mid-morning at the latest for an important meeting. He added that even though it was in the middle of the night, as a friend of long standing, he would be greatly obliged if the repairs to the car could be done without waiting. The garage mechanic immediately agreed, promptly lit the fire, and completed the welding of the cracked leaf of the damaged spring all to Idris' complete satisfaction, well before daylight. When Idris asked what he owed him, in spite of repeated requests, he just would not accept any payment. All that they said was that on this occasion it was Idris who had had a problem and they were most pleased to help; then adding that they were sure that if either of them faced a crisis of this sort, Idris' response would be no different! What wonderful people we have in our wonderful country! Before Idris left, the truck driver told him that the reason he was so insistent that Idris should get away from the village where he had stopped, was because it was notoriously dangerous. It was known that the police and a village gang there were well connected, and always on the look out for travellers like Idris in a well loaded car to be led to a lonely spot, beaten up and looted. What an escape he had, thanks to a very kind Samaritan.

For our accommodation, Idris had been allotted a rather small but quaint little cottage, called 'Dawnford', in Conoor, just a short distance from the Staff College in Wellington for the almost year's duration of the course at the Defence Services Staff College. The course, we found, kept Idris terribly busy. However, there were people from different nationalities attending the course who had their wives along too, and we had a delightful group in all our joint activities and at social gatherings. Idris not only joined in the hunts on horseback, but also on one occasion, rode in a regular Gymkhana race. There was great excitement when galloping to the winning post, he was declared second in the race. At the end of the course, I typed out all of Idris' final

papers, and was quite nervous doing so as I could only type with two fingers and took ages to do so. What I thoroughly enjoyed was taking part in plays and dance dramas on stage, and wearing my glamorous clothes. Once, however, I had to act the part of a batty English spinster. We also went on a delightful overnight safari to Bandipur, where we went through the forest on an elephant along with Idris' mother and our two children, Mariam and Asad. Not too far from our house, there was a green forested area where we could walk and pick some delicious red berries which we were told were a variety of wild raspberries and safe to eat.

At the end of the ten-month course, we undertook a very enjoyable trip in a hired station van. Sheila and Lakshman Katre with their son, as well as Major Eric Diaz, Major Gursharan Singh and a Colonel of the Nepalese Army, along with the two of us, our two children, our ayah and the chauffeur (12 in all) were able to sit — cramped, though, in a way enjoyable also — in the station van that we had hired. In the interest of economising by not having to hire accomodation en route, I remember so well, we used to drive at night (with a few sandwiches and coffee for sustenance) and do the sightseeing by day! Our trip first took us to Madurai, and then on to Cape Comorin, to Trivandrum, to the Periyar dam. Also included were visits to the fabulous Madurai and Suchindram temples, and, finally, to Trivandram, where we were all thrilled to be able to attend the midnight mass in a magnificent church on Christmas Eve.

After the course at the Defence Services Staff College at Wellington, Idris was posted as the Senior Air Staff and Administration Officer (or SAASO — as it was then named) to the Air Force Base Repair Depot at Kanpur under the command of the dynamic Air Vice Marshal Harjindar Singh. A third great joy in our lives was the birth of our third child, our son Asgar, who was born on 13th December 1958, in Hyderabad, where I was able to go during the period of our posting at Kanpur. He was born at Dr. Burgaonkar's clinic, with my beloved mother by my side. As he was born, we heard a sonic bang from one of the supersonic aircraft based at Hakimpet, which flew overhead as if to welcome him! He was a big healthy little boy, full of smiles and gurgles. Dr. Burgaonkar, most serious and strict and able, was my very fine and caring doctor for the birth of each of our three children. He loved to grow roses and would give them special nourishing injections. They grew beautifully and he was kind enough to bring me one every morning while I was there.

We stayed at the Kanpur Club for several months, as there was no vacant accommodation available for us at the Air Force Station. When driving around, trying to find a place for us to take on rent, I found that when I stopped to ask directions to places, the people on the road, as well as the policemen were charmingly polite, and would take great trouble in explaining the route to me. I was really most touched and impressed by their kindness. The people of Kanpur were, we found, certainly amongst the most helpful, courteous, warm and hospitable; and we really felt very privileged to have got to know some very fine families there. We soon moved into a house that fell vacant at the Air Force Station, a house which was similar to the one we had when we were at the Joint Services Wing in Dehra Dun. This house, however, had a kitchen garden also at the back and, most importantly, unlike our place in Dehra Dun, we had the luxury of running water and proper sanitation with a flush system for the toilets. The soil in this area was so fertile that in no time at all we had flowers growing in the front garden, and good vegetables in the kitchen garden.

CHAPTER 13

Hyderabad: Commanding Two Training Wings

In 1959, after barely a year in Kanpur, Idris got a call from (then Air Commodore, I think) Erlic Pinto at Air Headquarters, an officer whom Idris greatly admired. Those who knew him and his wife Verna had a very special feeling for both of them. One morning, he unexpectedly telephoned Idris and asked him if he would like to be posted to Hyderabad, in overall command of Begumpet and Hakimpet. This, he added, would be on promotion to the rank of Group Captain. If agreeable, he asked Idris how soon would he be able to move. This would be Idris' first command of a station and he jumped at it, jokingly responding (only because, Idris confessed to me later, he felt so very close to Erlic) "Will half an hour be OK, Sir?" Apart from the command of a Flying Station, we were thrilled at the prospect of being back in our hometown for a while, coupled with the prospect of being close to both our mothers and many others in both our families. Ultimately, an added bonus in a way was that, because there was no suitable accommodation for a Group Captain on the

Idris 1959, Hyderabad

camp, a spacious house with plenty of space for a nice garden had been hired for the previous Station Commander, and would now be allotted to us. The house, by a stroke of luck, we found, was also less than a kilometre from where Idris' mother lived. To top it all, since it was a spacious house with four bedrooms, we were able to have my mother move in with us. An added bonus was that her beautiful furniture greatly helped to enhance the elegance of our home there.

Idris, commanding the parade, in honour of the C-in-C Training Command,
Air Vice Marshal Ranjan Dutt on his first formal visit to the Air Force Station, Hyderabad, 1959

Air Force Station, Hyderabad, had two wings — TTW or the Transport Training Wing at Begumpet and, linked with it, of very great importance for all our twin and multi-engined squadrons, including our Canberras, the Navigation Training Unit, equipped with the good old sturdy Dakotas of World War II vintage. This is where Idris' headquarters was also located. At distant Hakimpet, a good 15-20 miles away, was the FTW (Fighter Training Wing), equipped mainly with Vampires, our Air Force's very first jet fighter aircraft. With Begumpet nearly 10 miles away from our home and Hakimpet another 15-20, Idris had to spend a lot of time on the road on most days. Official transport was not authorised, even for the Station Commander, except when travelling between work and home. Idris had his Vauxhall car, which he had bought when he was a bachelor, and used it also to participate in the compulsory games and sport at the Air Force Station for all in the afternoons. He believed it was important, as he told me, to take part in all the sports and games activities, as that brought him

closer to all ranks. In the process, it provided yet another tool to keep a finger on the pulse of the morale of his team of officers and men.

Dakota, Transport Training Wing, Begumpet

Vampire, Fighter Traing Wing, Hakimpet

At work, he was most thrilled at this opportunity of getting his fill of flying on fighters (the Vampire at Hakimpet) plus, having been only on fighters till then, it provided a golden opportunity to fly and qualify on the twin-engined Dakota. After he had done his solo flights and gained a few hours of experience, Idris started a weekly supply-dropping competition primarily to introduce an element of healthy competition and some excitement. For this, Idris got a target area with 2-3 circles marked at an unused corner of the airfield. This was possible as, for years, the airfield had been under the Air Force and only limited civil flying had, as a matter of policy, been permitted. Thereafter, once every week, all participating pilots — with Idris participating as often

as time allowed — would put a five-rupee note into the kitty to be handed over to the pilot who, on his practice supply run, dropped a chalk and sand-filled gunny bag closest to the bull's-eye. Happily, it proved a contest very keenly looked forward to, as the CO of TTW, the late Wing Commander George Joyce often told Idris.

I was involved in station welfare activities, but distances presented a major problem. The Station Commander's wife, even when promoting welfare activities, was not entitled to the use of Service transport. As a consequence, to travel to distant Hakimpet and Begumpet fairly frequently, I had to borrow my mother's little Morris Minor car.

A fancy dress party at the Officer's Mess, with Idris standing on the extreme right

Not only during games and sports, but at cricket matches also our personal interaction with the cadets proved very popular; and we also invited the cricket teams to join us at our home for informal dinners. One of the good cricketers was Pradip Biswas (son of the famous music composer Anil Biswas), who had a deep and lovely voice and enthralled everyone with his singing; in particular, Idris and I still remember when he sang "*Chaudhveen ka chand.*" It was a terrible tragedy when, one

day, he was killed in a flying accident — a truly shattering experience even for both of us, for with our close interaction, he with many other cadets, had become almost like members of our family.

Idris being the senior-most Air Force officer in Hyderabad, our social life, including entertainment at home, was pretty hectic, but all the officers, cadets, Non-Commissioned Officers (NCOs) and airmen, as well as their families, were very warm and friendly. Also, of course, we thoroughly enjoyed our get-togethers with our civilian and Service friends in Hyderabad. I still remember a 'cowboy barbecue' evening in fancy dress organised by a charming and enterprising officer 'Sikky' Sikand (retired, not long ago, as Air Marshal). The periodic graduation parades of the trainee cadets, organised at Begumpet, were also, in every respect, amongst the best that I have ever attended.

It was on this posting that Idris started what he called a Guest Night in honour of visiting dignitaries. At these, the Station Commander himself presided, instead of the PMC (President, Mess Committee), and the officers' wives would also be invited. In this aspect, it was distinct from the standard weekly/monthly compulsory formal Dining-in-Night which, in a sense, was traditionally a compulsory 'parade' which all officers had to attend in formal evening uniform with medals. At these Dining-in-Nights, the officers would be seated at the dining table strictly according to rank and protocol. Dinner was served by smartly-dressed waiters, also in uniform. Each course was served only when the Vice President who, for the occasion, sat at the down end of a long table, opposite to the President of the Dining-in-Night, signalled to the headwaiter to commence serving. Thereafter, every officer had to finish the course before the next course was allowed to be served. At the conclusion, a toast to the President of India (as the overall Commander of the three Services) would be announced by the Presiding Officer at the Dining-in-Night, and everyone would stand and comply.

It seemed very impressive, in fact, rather grand. However, I remember one of the officers telling me that there was an occasion when a recently posted officer, clearly unconcerned about the others, had a habit of chewing and chewing and chewing his food at each course. He continued eating long after all the others, including the President, had finished, and everyone else was impatiently awaiting the next course.

Even when the great inconvenience to others was tactfully brought to the officer's notice by friends, and then, seniors, he would say that from childhood he had been told that it was unhealthy to swallow food unless each mouthful was chewed thirty times! A stage ultimately came when this was brought to Idris' notice for his guidance. To everyone's delight (including the culprit) Idris ruled that the officer be allowed to absent himself from this 'parade' until he was posted elsewhere.

Idris had told me that from its birth, our Air Force had been following British traditions and, as a rule, ladies were always excluded from the Dining-in-Nights. Idris said that whilst these Dining-in-Nights should continue, on special occasions, such as a formal visit by a senior Commander or dignitary, he felt strongly that, as the majority of wives were always so helpful in so many station activities and, in a sense so much a part of the Air Force family, they should be included. However, since the Commander-in-Chief was coming on a formal inspection of the station, and it had never ever been done before in the Air Force, Idris thought it best to obtain the concurrence of Air Vice Marshal Ranjan Dutt (then Air Officer Commanding-in-Chief of Training Command), who spontaneously agreed to the Guest Nights. The wives were also pleased, as they too felt, even more, that they were members of the Air Force family, and this added a special meaning to their involvement and role in station life. Within months, we learnt that at most other stations also, beginning with Headquarters Training Command, for similar occasions, these Guest Nights became standard practice — in fact, the rule, rather than the exception.

Another popular and successful annual event that Idris started was the Best Homes Competition for the married, senior NCOs who had been provided Service accommodation. He had noticed that their domestic area, though generally clean, had little personal contribution towards its beautification, and felt that such a competition may well encourage them — and others — to make their home and areas more attractive. Idris asked me to head a small team of judges. It was a delight to see the enthusiasm of the participating families. Several of the wives came up with inexpensive little innovations, for instance, by utilising waste objects to make little basins in their garden, for flowers. Some had made cushion covers and even borders for matching window curtains from leftover scraps of cloth. Many of their little gardens were truly charming, and the families seemed to thoroughly enjoy the challenge of this competition. We found that they were also very appreciative of the opportunity it

provided to them for a much closer and more meaningful interaction with the wives of the officers. The experiment succeeded beyond expectations and became an event that the NCOs' families, each year, genuinely started looking forward to. So much so, in fact, that Idris held similar competitions when he later commanded 2 Wing; and again, at Bamrauli when he was in command of Central Air Command. In Nagpur, we also had an officers garden competition. At each place, in addition to the usual Air Force Wives Welfare get-togethers, this kind of activity gave me a truly welcome opportunity for greater interaction not only with the families but also the officers and men. In a sense, this made the whole station one big united family. Both of us have happy memories of our posting in Hyderabad. Barely 18 months later, Idris was told that he had been nominated as the Air Attaché to our Embassy in the United States and we, with the children, would have to move to Washington at very short notice.

CHAPTER 14

Uncharted Waters

In the third week of December 1960, we left by sea from Bombay for Idris' next assignment as Air Attaché at the Embassy of India in Washington D.C. We reached the United States of America a month later in January 1961. We arrived in the freezing cold weather of Washington just before the inauguration of the charismatic John F. Kennedy as the President of the United States, and were even invited to two of the special inaugural functions which was really quite a unique and enjoyable experience. After a short stay at a hotel, we moved into a furnished three-bedroom house with a large, airy attic and basement, as well as a garden all around. It was in the charming Chevy Chase area, and we chose to live there because it had very good schools nearby. From our home in Chevy Chase, we looked across to a narrow part of the lovely woods of Rock Creek Park. Many of the Air Attachés of other countries of the same rank and status as ours, however, lived further out, including in Bethesda and Virginia, where the homes were larger, much more modern, and far better furnished. These were well beyond our government's approved financial rental ceiling for Defence Service Attachés. However, Chevy Chase was also a very beautiful area and also much closer to the Embassies. Actually, we found a few very good schools — naturally, of very great importance to us — which were very conveniently close to the house we finally selected. To top it all, our next-door neighbours too were warm, friendly and most helpful. Altogether, Idris and I and, importantly, the children, have very happy memories of the years we spent there. With the carpets, antiques and *objets d'art* that we had brought with us, our colonial style home became quite charming.The garden both at the back and front added to this.

The Americans we were fortunate to get to know were, without exception, very friendly and hospitable. For Idris, the initial challenge was to first recognise and then, selectively, establish warm working relationships with important elements of the United States Air Force, the State Department (their Ministry of External Affairs) and also the Attachés of countries of importance to India. Soon he found that this was not difficult at all as, almost without exception, all of them were warm and welcoming. The most important were, of course, those in the Foreign Liaison Department located conveniently at the Pentagon. In particular, I recall, General Hall, his wife and lovely daughter, Colonel and Mrs. Dallam, and Colonel and Mrs. Burrit, all no less warm, welcoming and friendly, were excellent and most helpful. Idris found them particularly so after the Chinese aggression of 1962, when our country sought the assistance of the United States, in particular the US Air Force and the State Department. Others who were also both charming and friendly were Mike Kettlehut and Charles and Elizabeth Rousek. The latter had lived in Hyderabad as a child, as her father who was a senior British Army officer was stationed there at the time.

Our home at Chevy Chase in Washington D.C., 1961 —65

When Idris travelled as the Air Attaché mainly to different Air Force bases and a few defence aircraft manufacturing establishments, he, along with the other Air Attachés, was invariably included in specially arranged tours to various bases, which gave him an excellent opportunity also of making some very good friends. After about a year, Idris was promoted to the rank of Air Commodore, which gave him the required seniority. This, in the United States, was important as the rank of Group Captain was considered equivalent to a US Air Force (USAF) Colonel, whereas Air Commodore was equivalent to their Brigadier General. This placed him in the category of "General", which made a major difference when working with those at higher levels.

Idris found the Air Attachés' periodic group visits to the USAF bases not just enjoyable but professionally valuable. It was in 1962, on the way to Hawaii that the group halted at a USAF base in California. In the evening, while they were all dining before an early morning departure to Hawaii, Idris received, much to his surprise, an urgent call from Delhi. It was from his Chief, Air Marshal Aspy Engineer, telling him that the Chinese had invaded, and he should immediately return to Washington to be available at very short notice, to accomplish assigned tasks. Idris took the very first flight back to Washington and was back at his desk the following morning.

Within a few days of receiving the news of the Chinese attack, at about eight in the evening, Idris got a call from our Ambassador, Mr. B.K. Nehru, asking him to come urgently to his office. On arrival (Idris told me long after), he was shown a top secret communication from Delhi at the highest level, asking the Ambassador to meet President Kennedy as soon as possible and convey to him our request to make available to us in India a squadron of heavy bombers and several squadrons of fighter aircraft as well as air defence ground radar units. The Ambassador sought Idris' views. Apparently, Idris' immediate response was that the request had some very serious implications. For instance, since the Indian Air Force (IAF) had certainly neither air nor ground crew qualified on such aircraft, or ground equipment such as their radar, the request, if conveyed to the President, would imply our government's acceptance at the highest level, of sizeable elements of the USAF, manned and operated by American crews, operating with, and in support of, our own defence forces. Besides, if this was not so, the conversion of IAF air and ground crews to become fully operational would clearly take a very long time, and, therefore, would be of little or no value to stem the Chinese attack. In view of these implications, Idris suggested that it may be prudent

for the Ambassador to make sure that these factors had indeed been considered and cleared at the Prime Minister's level before he conveyed our government's request to the US President. To Idris' knowledge, nothing more was heard specifically on this!

The months immediately following the Chinese invasion of 1962 were, as expected, particularly busy; but, for Idris, he always said, not just challenging but most stimulating. Immediately following the then Defence Minister Mr. Krishna Menon's visit (though brilliant, he was not the easiest person to work with), Idris was asked to fly urgently to Canada. He was asked to go first to Toronto, to expedite the deliveries of the initial batch of Caribou medium transport aircraft that had earlier been ordered from de Havillands, and were now very urgently required for our Eastern Sector where the main fighting was taking place. Immediately thereafter, Idris went from Toronto, straight to Ottawa, to the Canadian Air Force Headquarters. It was to ask them if they could spare a few Harvard Trainer aircraft (a type in use with us also) to enable us, in India, to accelerate the training of our pilots. The Canadian Air Force, at all levels, Idris had often told me, was always, always warm, welcoming and helpful. In view of the extreme urgency, Idris sought, and immediately obtained, a meeting with their Air Chief. After Idris explained our need, stressing also the urgency, the Canadian Air Chief, on the intercom, immediately called the head of their Maintenance and Supply Department and asked him how many Harvard Trainers they had in storage. There were, perhaps, around twenty, he was told. Idris was, understandably (!), bowled over when the Chief directed their Maintenance head, in Idris' presence, to fully service all of them on top priority, and arrange on an emergency basis, for their delivery to our Air Force, in close consultation with our High Commission. No wonder, Idris always spoke so very warmly about the Canadian Air Force!

Idris and I often talk about our years in Washington, especially about our warm, brilliant and highly respected (late) Ambassador B.K. Nehru and Fori, his wonderful wife, who now lives mostly in Kasauli. She is now, even at the age of 102, vibrant and full of life. Mrs. Nehru kept the Embassy residence beautiful and elegant, and was a great example in every way for all of us to try and emulate. Her warm and loving nature has been for us, both deeply treasured and precious. A quite unique event during that period, I know, made a deep impression on Idris, and, till now, he has always spoken of the Ambassador's personal involvement in it, with both pride and admiration.

Whilst in Washington, more than once, I remember Idris saying that, sadly, very sadly, it needed the Chinese invasion to wake up our government to the long-felt need of modernising and strengthening our armed forces, our Air Force in particular. For instance, Idris always stressed the urgent need for a substantially stronger and state-of-the-art fighter force. When the Chinese struck, with John F. Kennedy as President who was deeply sympathetic, the transparently strong desire of the higher echelons of the US government to help India was becoming well known in India also. Idris knew that Air Headquarters had rightly decided that top priority should be given to strengthening the fighter element. A request for the relatively very high performance Lockheed F-104 Starfighter was imminent. This, he felt the US government could not object to, as it had already been supplied to Pakistan as part of their military aid programme. Idris briefed the Ambassador accordingly. During the discussion, when the Ambassador mentioned, more than once, that our economy was already under very great strain, Idris left with a feeling that our Ambassador would — and understandably so — be less than enthusiastic! But Idris was also convinced that for the country's defence, a more effective fighter force was not only inescapable but most urgent, and it was his duty to ensure that the Ambassador, particularly, lent his full support.

Therefore, having had the good fortune to have developed a close and warm relationship with him, Idris asked for another meeting with the Ambassador to get the ball rolling. This, happily, was arranged within the hour. The warm and informal way that Mr. Nehru received Idris greatly encouraged him, and, as an introduction, he again stressed the importance and extreme urgency of upgrading our fighter force and the likely official request. Encouraged by his sympathy and understanding, conveyed eloquently by a warm smile, Idris took courage in both hands and, as a starter, asked the Ambassador if he would like a flight in the Starfighter. To Idris' delight, the Ambassador, with great excitement written on his face, asked Idris, "Do you really think you can arrange it?" "Of course, Sir", Idris replied, "I'll ask Lockheed to position a dual-controlled aircraft with an experienced test pilot at the Air Force base in a couple of days, and, at your convenience, but preferably fairly early in the morning — say at about nine — when the weather is pleasant, we could drive down and still be back well before lunch". "How wonderful Idris, please go ahead; but remember, anything before lunch is indecent (!), so let's make it two in the afternoon." This was done.

At the airfield, after being strapped into his seat, before starting the engine and taxiing out, the test pilot, in accordance with normal Air Force procedures, first fully briefed the Ambassador on the various possible in-flight emergencies that could arise, such as engine malfunction or total failure, fire in the air, etc. This was something about which the Ambassador later commented privately to Idris, that had he known all this earlier, he might have thought twice before agreeing to Idris' suggestion so readily!

Ambassador Nehru in the Lockheed F-104 Starfighter

The pilot then taxied out and took off. Once airborne, the pilot told Idris later, he gained height, did some steep turns, dives and pull-outs and, finally, after a supersonic run, allowed the Ambassador to briefly handle the controls. The pilot then asked him if he could return to base. "What", apparently was Mr. Nehru's response, "no aerobatics?" Delighted, the pilot climbed again, did some rolls and loops and then finally, landed back. Luckily, Idris was able to get a wonderful picture of the Ambassador getting out of the cockpit, a picture of sheer delight on our Ambassador's happy and smiling face! Then and there, Idris decided that in private personal conversations, he would henceforth, refer to him as the world's one and only Supersonic Ambassador!

Above all though, Idris always concluded his narration by saying that what impressed him no less was the exchange they had in the car on their way back, as it gave him a measure of the Ambassador's qualities of head and heart. "Idris," he had apparently asked, "how much fuel do you think we consumed in that 20-minute flight?" Idris did a rapid but very rough calculation and told him, to which his response was, "You know, Idris, that amount of fuel could have provided light in an Indian village for a whole year". He was quick to add, "I am not suggesting, that you shouldn't buy good aircraft; but always, always, keep in mind the effect of each such step on the country's economy". After a minute or two of silence, Idris' response was, "Sir, you are absolutely right; but I'm sure you will also agree that when our pilots go into battle, on every single flight, it is not only a matter of successful accomplishment of the task assigned, but, at the very personal level, it could be a matter of life and death; and none in our country, I'm sure, would want them to be flying aircraft that we knew beforehand, were clearly inferior in comparison to those of our adversary." The Ambassador gave a gentle smile in response and nodded "I know, Idris, don't ever think we would".

Within days, our Embassy was officially informed that an Air Force delegation, led by Mr. Y.B. Chavan who had taken over from Mr. Krishna Menon as Defence Minister, and included Air Vice Marshal Arjan Singh, the Vice Chief of the Air Staff as the senior Air Force negotiator and spokesman, would soon be arriving in Washington. The purpose of the delegation was to convince the United States authorities of our most urgent need of, among other requirements, the Lockheed F-104 Starfighter — and nothing less.

Idris had done other preparatory work also, and was able to warn our delegation, immediately on their arrival, of the reluctance of the Americans to accede to our request, and their intention to do their utmost to get us to agree to accept a couple of squadrons of a relatively obsolescent type of aircraft, even, if it became necessary, at throwaway prices. These aircraft were originally with their Navy, Idris has told me, but at that time were with their National Air Guard, which was, in a sense, their second line of defence. In fact, as a special gesture (perhaps to soften the anticipated reluctance of our Air Force?), the USAF had arranged a special flight on the type for Air Vice Marshal Arjan Singh the day before the first formal meeting. When Air Vice Marshal Arjan Singh and Idris were together briefly soon after the flight, and Idris

asked him how he enjoyed the flight, he told Idris with his seldom-absent charming, almost naughty (!) smile, "Wonderful! But I wasn't born yesterday; and you had, fortunately, already warned me of what lay behind."

Flight on the aircraft arranged for Air-Vice Marshal Arjan Singh. Courtesy US Air Force

At the first formal meeting at the ministerial level the following morning (Idris told me much, much later, according to my little diary of occasional notes — long since destroyed), a USAF specialist team tried its very best, by an elaborate and detailed presentation, to convince our delegation that what they had in mind was really the best option. Two reasons were stressed. Firstly, in view of the great urgency, a full two squadrons' worth of aircraft could be made available almost immediately. Secondly, not unimportant, at a fraction of the normal cost; in fact, finally offered at no cost. Fortunately, Idris told me, Air Vice Marshal Arjan Singh made an excellent and telling presentation in which he, almost mathematically, proved that anything less than

the F-104 would be of little value. On completion of his masterly presentation, one could sense that even the USAF specialists felt quite cornered. Soon thereafter, it was decided that all would meet again the following day to take a final view.

The following morning, Mr. Chavan met the Air Marshal and Idris at the Embassy for an informal exchange of views before the formal meeting with the US authorities. Both of them were dreading that, in view of the seemingly very attractive offer to Mr. Chavan — the prospect of returning to India, after his very first challenging task as Defence Minister, with two full squadrons of fighters free of cost — he might, understandably, be tempted to accept the American offer. Perhaps sensing their concern, to their delight and great relief, Mr. Chavan opened the discussion with the words. "Don't worry, Air Marshal, I will never accept anything of which you do not fully approve!"

At the formal meeting, the Air Marshal, with transparent confidence, patiently but tellingly, Idris told me, once again, explained the reasons why the fighter suggested by the Americans would, in view of the capability of the Chinese, be of little or no value — reasons which the USAF team had found impossible to counter. The American side then said that since a mutually acceptable solution of the fighter requirement did not seem possible immediately, they would be more than willing to give us, without delay, to begin with, at least a few C-130 transport aircraft against the eight we had requested; and also equip some of our ground radar units, as we also requested. Sadly, though, in the context of the Chinese aggression, no C-130s were ultimately sold to us. However, when our Air Chief finally left, Idris and I felt so very proud and happy that he left behind such a richly deserved and wonderful image of not only his personality and social graces — always smiling, always warm and friendly — but also, indeed no less, his outstanding professional competence.

Much later, President Radhakrishnan came on a formal visit to the United States, and President and Mrs. Kennedy attended a reception held in his honour at the residence (at Malcomb Street in Washington D.C.) of our Ambassador, Mr. B.K. Nehru, and his very charming wife, Shobha or 'Fori', as our families knew her. Mrs. Kennedy looked even more attractive in real life than she did in her photographs. President Kennedy was also delightfully informal and charming, and very soon was able to put others at ease. President Kennedy, who spoke so brilliantly, made a great

impact in many parts of the world. I have repeated his inspiring phrases so often, especially the one which says, "Ask not what your country can do for you, but what you can do for your country."

Tragically, very tragically, later the President was, as we all know, shot in Texas and died too young and too soon. Idris went with Ambassador B.K. Nehru to the airfield and was present when Mrs. Kennedy returned to Washington D.C with the President's body. Idris mentioned to me that she was still wearing the dress she had on when he was shot in the morning, all smeared with blood when she had cradled his head in her lap — a very moving and tragic figure. Strangely, I had been watching the one o' clock news on television that afternoon and had seen the whole tragic scene live when he was shot by a sniper. What a terribly shattering way for this brilliant leader to go. Till today, I can still feel his loss.

An interesting event took place in connection with the Air Attaché of a country that was friendly with India, but did not have close ties with the United States, to the extent of being excluded from the officially organised annual Military Attachés' tours. These were primarily to Army, Navy and Air Force establishments, as well as one or two places of national importance. The Defence Attaché of this particular country called up Idris one day and invited him to have lunch with him at a restaurant. When they met and sat down at a table that had been reserved, the host requested the headwaiter to move them to another table which was still not occupied. It immediately occurred to Idris, he told me later, that his host may have done so suspecting that the table which had been reserved for them may have been bugged. During the meal, after exchanging pleasantries, he asked Idris if he could help in getting him some information during the forthcoming Attachés' tour. Specifically, he said, he was keen to know about two aspects concerning the US Air Force: firstly, the maximum bomb loads of their heavy bomber aircraft, and, secondly, the accident rate of the US Air Force.

Idris was prepared and, fully conscious that should he give in and agree to pursue this; being accredited to the US and its Air Force, it would be morally totally wrong. Also, as the US intelligence would, sooner or later, get to know, it would almost certainly do irreparable damage to his standing with the USAF and the government of that country and our own. At the same time, since India's relations with the country of his luncheon host were cordial, Idris did not want to

just bluntly turn down the request. Carefully choosing his words, he responded by saying that he believed that, in principle, it would be wrong for him to pursue this request. In explanation he added, "Please imagine that if I were accredited as Air Attaché to your country, and the Attaché of a third country had asked me to surreptitiously obtain and pass on such information, I am sure you yourself would feel that it was a betrayal of trust." Fortunately, Idris said, his host smiled and nodded in agreement. However, not to dismiss the request entirely, Idris went on to say that as far as the bomb load was concerned, it was public knowledge that it carried a very heavy load, and, therefore, whether the exact figure was a little more or less, may make little material difference. Idris was again greatly relieved to see him nod in acceptance. Finally, Idris addressed his host's question about the accident rate of the US Air Force, saying that this was a common concern of all Air Forces and he felt sure that the US Air Force would be more than willing to exchange this information if his host, whom he knew was greatly respected, were to make a request directly to them, while offering to share similar information about his own Air Force with them. Idris was greatly relieved to find that his host did not take his suggestion amiss, and responded by saying that the suggestion was certainly worth thinking about.

The very next day, Idris went to meet the senior-most officer of the USAF Foreign Liaison Department, whom he knew well; in fact, we both had a very cordial relationship with him and with his wife. Idris briefed him fully of his lunch meeting, adding only that, perhaps, an exchange of information between the USAF and the other country's Air Force on accident rates could be of mutual advantage without compromising the security of either nation. The US Air Force Colonel, after expressing his deep appreciation of Idris' frank and spontaneous action, to Idris' delight, added that he would, very soon, not only invite the senior Attaché of the country concerned to one of their air bases for a full and free discussion, but ensure that he was received as a very special invitee. Sadly, within a week, there was an unfortunate episode between the two countries which made it impossible for this initiative to be pursued. But clearly, Idris' suggestion, particularly his reporting of the conversation to the USAF, made a deep impression on them and perhaps even contributed to their decision, when his term as Air Attaché was over, to confer on him their special Medal of Honour. However, Idris was informed by our government of its policy against the acceptance

of foreign awards by its Defence Services Attachés, and this was tactfully conveyed to the USAF at the Pentagon.

General Mckee, the US Air Force's Vice Chief, was also warm and friendly and so was his wife, as we found when they had dinner with us one evening. Some months later, when General Mckee was to visit India, at a reception that he hosted, he asked Idris and me what he could bring back for us from India. Without a second's hesitation, we both said that we would love to have some 'paan' or betel leaves, explaining, as best as we could, what they were! We were both very touched when the Vice Chief spontaneously agreed and, sure enough, he brought back a goodly bundle of them. A few days later, however, when we met him again at a reception, he said that they could not be delivered to us because the US Customs insisted on burning the lot, as any fresh fruit, leaves or vegetables were not allowed to be brought inside the country! Neither he nor we had realised this when we asked for them.

Another rather heartwarming memory is when, after the Chinese invasion of 1962, the USAF, at Idris' request, without hesitation, offered a number of vacancies at their premier fighter pilot training establishment, for further upgrading the specialised skills of some of our fighter pilots. Whilst we were still there, our Air Force sent several batches over a year or two. It was a matter of great pride and joy for all of us at the Indian Embassy, when the local papers, on more than one occasion, prominently reported that our trainee officers passed as the toppers of their respective courses, making actual headlines in the local newspapers, cuttings of which Idris invariably personally showed the Ambassador, our much admired (late) B.K. Nehru.

Perhaps about two years later, Air Marshal Arjan Singh, by then the Chief of the Air Staff, came again to the States, this time in response to a formal invitation from the head of the USAF, General Curtis LeMay of World War II fame when he had been their Chief of Bomber Command. Deservedly, he was highly regarded for the USAF's major bomber operations against the Germans, but was also well known for being a man of few words. During a formal visit, when Air Marshal Arjan Singh, our Air Chief by then, accompanied by Idris, called on him, they both found him warm and welcoming! The latter was generally considered hard to converse with, yet according to Idris, he not only met them most cordially, but directed that any special visits that our Chief may want to make, even to the USAF's highly classified formations, should be arranged.

During our Chief's visit, at his request, a formal and very high-level meeting was also arranged at which Mr. McNamara led the American side, and ours was led by our outstanding Ambassador, Mr. B.K. Nehru. At the meeting, Air Marshal Arjan Singh formally requested the US government, not for the F-104 Starfighter (which they had earlier not agreed to), but for a couple of squadrons of the Northrop F-5 to help modernise our ground attack capability. The American government had, of course been earlier advised that such a request would be made. It was Mr. McNamara who immediately responded by saying that in their view also it was an excellent choice that we had made, and, in principle, it was entirely acceptable to them. He then suggested that initially, a couple of squadrons could, perhaps be considered. He finally enquired when and where (in India or the US) we would prefer the announcement to be made. Our Ambassador, warmly thanking Mr. McNamara, suggested that since he was due to visit Delhi very shortly, a final decision could perhaps be slightly delayed. This was spontaneously accepted. Within a day or two, Air Marshal Arjan Singh returned to India. Sadly, very sadly, however, as far as our request for the F-5 aircraft was concerned, so readily accepted at the level of the US Defence Minister himself, they never came because, within months, the Gulf of Kutch armed exchanges with Pakistan took place, and the Americans said that until peace was restored, further movement could not, very unfortunately, be considered.

Idris after flying the Northrop F-5 in California

We were deeply involved in semi-official activities also, and entertained a great deal as well. Whilst in Washington, Mrs. Nehru organised a lovely costume show of Indian costumes. Inspired by this, I also organised several rather elaborate Indian costume and cultural shows at US air bases, which were greatly appreciated. In fact, the audience would really get interested and asked a lot of questions about different aspects of our country. They also enjoyed the dances performed and customs depicted by our ladies. I was requested by Mrs. Ross Adair, the wife of Senator Adair, to travel to different states to speak about Indian women, customs and faiths at universities, special institutions and clubs, which I then undertook. There were five of us from different countries who were invited to speak about our own countries. It was very interesting as I was asked to stay with several American families in the various states, and while I spoke about our own, I also learnt a great deal about the United States and its people as well. One year, I was also asked to be in charge of a particular section of the Cherry Blossom Parade, which was a great event organised in Washington in spring every year. Ambassador B.K. Nehru then asked me to lecture in Philadelphia for a university and a Rotary group,

Friends dressed in my sarees at a dinner at our home

Mariam filmed on television performing a Manipuri dance in Washington DC

Asad in a school play dressed as a cowboy

Asgar wearing an achkan

which was a great success as I also organised a cultural show in which a group of talented Indian women and children from Washington participated. This was soon after the Chinese invasion in 1962. At the end of the show, what was very touching was that members of the vast audience donated huge packages of warm clothes which they had brought along to be sent to India for the people who needed them in the areas of conflict and suffering. Mrs. Nehru arranged for these to be sent on.

The Americans were hospitable too, and we were invited by them to their homes. This involves a great deal of personal involvement for the hostess, and often for the host too. Mrs. Dean Rusk, the wife of the Secretary of State, was another very warm and friendly lady. On one occasion, at a lunch party at her house for a few ladies, she served a superb dessert which I liked and praised, and she promptly showed us how to make it. At our own dinner parties at our home, I would often lay out sarees for the ladies to wear, and then they looked very lovely in them. I was also very impressed on an occasion when I was invited to tea by Mrs. Curtiss LeMay, the wife of the US Air Chief, where I noticed something quite unusual. This was the paving outside the entrance to her house which was very attractive. She told me that she had designed and laid it out all on her own.

Mariam, Asad and Asgar liked their teachers a great deal, and were happily involved in their school activities. Mariam was learning our Bharata Natyam and Manipuri dance forms from an Indian teacher in Washington too, and was asked to perform at cultural functions and even on television. Her Manipuri costume was actually stitched and made by me, as these were not available there. Mariam was full of caring for her brothers, even buttering their toasts at breakfast, as she blossomed into a lovely girl. All three of the children soon had good friends in their schools. Asad was selected by the school to help with children crossing safely from the road outside the school, for which he had to wear a special belt and badge for the task, and looked a smart little six-year-old. Asgar, who was only just two years old when we got to Washington, joined play school and made lots of friends and became very attached to his teacher. Through their growing years, wherever we were, Asad and Asgar were such a happy duo, always full of *joie de vivre*. Asad was full of jokes, some not always funny, with Asgar, without fail, almost dutifully laughing away at them. Actually, he kept us all laughing.

Idris and family on a Sunday outing

Great learning experiences for the children and for us personally, during our five-year stay in the United States, were the trips by car, in our Chevrolet Bel Air. Idris organised these each year during the children's summer vacations, when he could also take 2-3 weeks leave. These were long trips by road throughout the length and breadth of the United States and Canada. On our very first trip, up north right into Montreal and Quebec and then to Ottawa, our first halt was in New York. The very first day that we spent there turned out to be most nerve-shattering and it was entirely my fault. We had taken the children to the World's Fair. After several hours there, I discovered that I had misplaced my handbag. In it were the car keys, but far worse was that it also contained our passports, money, and all my lovely jewellery, now gone! A frantic search began and we went round all over. Suddenly, a friendly looking American couple came up to us and asked if we were looking for something. When we explained our predicament, they advised us to check at the Lost Property Office, even guiding us to it — bless them. Believe it or not, as we approached, a smiling police officer came towards us holding up my bag in his hand, asking, "Is this what you were

looking for, Madam?" God had really come to our rescue. Amazingly, everything was there: my jewellery, our passports, the keys for the car and all our money — with not a penny missing!

I must mention that, on our return to Washington after the holiday, the very first thing that Idris did at his office, was to write a letter to the concerned authority in New York sharing our deep appreciation and heartfelt gratitude. A great lesson learnt by me; and one that we will never forget. All told, each trip was well over 5,000 miles. Idris drove most of the time, although occasionally I also took over. Throughout, our three children were as good as gold; occasionally, when a little tired, they slept at the back; with Mariam stretched out on the seat, Asad on the floor and Asgar (maybe, still just two years old) up behind the back seat against the rear glass! Altogether, in three years, three truly wonderful trips — North, East and South — not only most enjoyable, but an education for all of us. These gave us a great opportunity to occasionally meet many warmhearted and welcoming American families, living in small towns where life, we soon realised, is very different from the big cities — as, of course, it is in every country. With Idris' enthusiastic photography, the highlights are preserved for posterity on films that we immensely enjoy seeing even now.

On our Air Chief's second visit to the United States, he had asked Idris if he would be prepared to stay on for another two years. Idris, of course, happily agreed, mentioning only that, for the sake of the children, if at all possible, his transfer should be at the end of the academic year — a request which the Chief spontaneously accepted — such a wonderful Chief to have. So, in June 1965, we finally set course back home.

CHAPTER 15

Air Defence of Our Eastern Frontiers

From the US, we returned to India in the summer of 1965, and went on to Shillong where Idris had been appointed as the first Air Defence Commander for the Eastern Theatre. His office was located in Upper Shillong, conveniently alongside the Headquarters of Eastern Air Command. The threat from (then) East Pakistan was limited, but so also were the resources available in terms of fighters, radar and, of critical importance also, a very high-grade communication network. It was a challenge and Idris, from day one, had a really busy schedule. But I could see that he relished it and soon began to hold a series of exercises to test the network that he and his team had developed. They all felt that a very adequate system of security from air attack was in place. In fact, till the end of the conflict, the Pakistan Air Force (PAF) got through only once when a section of two PAF arcraft slipped in at a very low level and attacked our airfield at Bagdogra. In any case, Bagdogra being very close to the border was, we all knew, very vulnerable. Tragically, one of our Vampires that was taxiing out just then was destroyed; and terribly tragically, the pilot was killed. His young wife was in Shillong and was completely broken when she was told. She faced her future thereafter with admirable courage.

When we first arrived in Shillong, we were put up at the Military Inspection Bungalow, which was on a hill with a magnificent view. However, because of the war and its fallout even in the Eastern Sector, limited though it was, the people in general seemed very nervous, and some were quite scared. I still remember that Idris once got a call in the middle of the night to say that they had sighted paratroopers being air-dropped in some remote areas. This contingency had, however, been anticipated and

the drill planned and well-rehearsed. Immediate action ensued, but after an extensive and thorough search, fortunately, and to everyone's relief, it was found that it had been a figment of the caller's imagination! Similar calls came several times!

Air Marshal Malse, who was the Air Officer commanding the Eastern Air Command, felt that the location of our Inspection Bungalow was rather insecure, especially as Idris and the children, at dawn and again at dusk, had quite a long walk, up and down the hillock and also to and from the main road. Very thoughtfully, therefore, he shifted us to the Government Circuit House which was located right in the town. This kind of caring and consideration was deeply touching and, through the years, has done much to make the Air Force into a real family. Shillong was a beautiful place with many very friendly and warm people. Very soon, we had friends amongst the Assamese and Khasi people, and we established a very warm and cordial relationship with them. Fortunately, there were excellent schools, and our three children were soon admitted, and with their usual hard work, and adaptable ability they got adjusted again to the Indian school system.

Air Marshal Malse also did much to establish cordial relationships with the Air Force officers and their families. Not infrequently, he arranged picnics on Sunday mornings where his own family as well as the others got together, having idli and dosa breakfasts (!) and playing games together. As a consequence, the children also made friends with each other. Some of the picnics were arranged in a valley between hills quite close to the Air Force Station. Apart from his Air Force tasks and responsibilities, Air Marshal Malse, we soon found, had wide vision and many interests. Very soon, he and Mr. Bir Kapoor, the senior civil engineering official of the state government, realised that, if they could manage to build a dam across a narrow part of the valley, they could create a deep lake with plenty of water of which there was a great shortage, particularly in the area. Happily, the dam, under the watchful eye of Air Marshal Malse, got constructed in record time. The Malse Dam, as we called it, with the relatively small but perennial stream, coupled with the heavy monsoon rains, was full within the year. On Idris' suggestion, pedal boats were procured, and a good selection of fish put into the lake, and very soon, it became a delightful place for recreation as well. As the final touch, at a carefully selected spot with a grand view of the lake, a lovely cottage was added for official visitors.

Air Marshal Ranjan Dutt took over command of Eastern Air Command (EAC) from Air Marshal Malse after about a year into our stay in Shillong. Idris was appointed as the Senior Air Staff Officer and he visited the stations of Eastern Command which included Hashimara, Jorhat, Tezpur and Gauhati. When he went to the Air Force Stations with Air Marshal Dutt, he returned full of praise and admiration for the Air Marshal's most inspiring speeches at each station.

A beautiful place near Shillong was the Barapani Lake. One of our generous local friends, an elderly lady, was so kind and generous that she told Idris that since we loved the area so much, she would be happy to gift him a plot of land which she owned at Barapani. One would not have accepted this, but in any case, Idris would have had to decline as in Meghalaya, non-locals were not authorised to own property there. But we were both deeply touched by this remarkable and most generous gesture of the lady. In fact, we found that the people of the region were really exceptionally warm, and very special.

Idris being presented a Guard of Honour in Nagaland

In December 1966, at very short notice, Idris was posted out. He was to join the National Defence College course in Delhi which was starting within a few weeks, early in January 1967. Though considered a very highly prized opportunity, all of us, including the children, were very saddened. As the notice was very short, we immediately got down to packing once again, and planned our move.

Idris on a visit to a fighter station in Eastern Air Command

Since we would need our car on arrival in Delhi, Idris decided that there was no option but for all of us to drive all the way from Shillong to Delhi, and keep only the absolute essentials to take with us in our car. The main baggage was sent by train, hoping that it would reach us soon after our arrival in Delhi. The drive, though very long, with several night stops on the way, was an adventure — and good fun too. Along the way, for instance, we spent a night in beautiful Sikkim over New Year's Eve, staying overnight with Mr. Coelho the High Commissioner of India, and his very warm and charming wife who had insisted that we should do so! They were really perfect hosts and had a wonderful party for the New Year at their official residence, a beautiful old mansion built by the British much earlier. The guests included the Chogyal and his lovely wife Hope. Idris took some wonderful photographs of the flowers in their

garden and, at my request, portraits of some of the lovely children and elderly local people.

Nearer Delhi, we had to make a transit halt which we had planned should be at Benares. The highlight of this was a pre-dawn visit to the banks of the Ganges. With the gentle sound of the temple bells, watching the sun gently rising up through the mist was not only an unforgettable sight, but also a unique experience. Idris managed to take some wonderful photographs, one of which was asked for and used for the opening page of a book, *Benares* by the author Pierre Toutain. This was when we were in Paris some years later, during the Festival of India in France in mid-1985. A day before reaching Delhi, Asad, sadly, developed fever, and by the time we reached Delhi, we found that he had a severe attack of mumps. On arrival in Delhi, we went straight to the flat allotted to us at Raksha Bhavan, where officers attending the National Defence College course were accommodated for the duration of the course. We were fortunate to be given a large and airy three-bedroom apartment overlooking the grassy expanse of Rajpath, which was also near the old Central Vista Air Force Officers Mess. On every new posting, we always carried a small trunk packed with basic cooking items, with some crockery and food grains like rice, dal and spices, tea and sugar, so that we could cook our first meal soon after arrival. Idris had a kerosene stove, primarily for our picnics, but which came in very handy for such occasions as well. There was basic furniture in the flat and we had brought towels, sheets, blankets and clothes with us, so we could have baths. Using the stove, I cooked a simple dinner, and then all of us went off to sleep.

Raksha Bhavan was a good place for all the children of the numerous officers who stayed there while attending the National Defence College, and the families also soon got to know each other well. Conveniently, household staff also came to the apartments seeking jobs as soon as the officers with their families arrived! Asad's mumps improved, but then I developed mumps! The result was that I was unable to go on the rounds of the good schools, as I did on each posting. When leaving Shillong, we had been assured that the boys, who had been good students, would definitely be taken in by St. Columba's in Delhi. But Idris who, because of my illness, took them, soon discovered that this was not so. In fact, they said that they had a very long list of prior applicants and there was no chance of our requests being even considered, which was a big setback. Then when we learnt of another excellent school, St. Xavier's, as I was still down with mumps, Idris had to undertake this task too at a time when

he was, in fact, due to start his course at the National Defence College. Fortunately, Father Kunnunkal was the Principal, a most learned and impressive Jesuit and dedicated educationist. Sensing Idris' predicament, he agreed to allow the children to sit for tests. Happily, they did well, and were, to Idris' and the boys' relief and delight, admitted. Later, I took Mariam to the Carmel Convent which, to our great relief, also agreed to take her in.

Although we had all been sad at having to leave Shillong, Idris mentioned to me that the National Defence College (NDC) course was and perhaps still is, the most important and sought after course for officers in the final years of their career in the Defence Services. He was sure that it would prove to be of great value to him also, which he later told me it certainly proved to be. Central to the one-year course, Idris found, was not only being privileged to hear an address by the Prime Minister himself, but also by Ministers and/or top officials of key Ministries of the central government, the Defence Services, and of industry. Often, what proved to be of special interest, Idris told me, were the question-and-answer sessions that followed the lectures and talks.

An unexpected but most valuable bonus, apparently, was that in addition to our senior defence officers, there was also a small sprinkling of senior officers of the civil services and of the paramilitary forces, plus just a few senior Defence Services officers of other countries, all studying together. Those attending the course were grouped in intimately placed 'syndicates' of about eight, which provided a truly unique opportunity to establish close friendships (so important an element in ensuring even greater harmony between the different arms of the Defence Services).The daily interaction with seniors of the other Defence Services also resulted in much better knowledge and understanding of the challenges, the capabilities and the constraints of the sister Services.

In 1968, after the completion of the National Defence College course, the (then) Air Chief Marshal Arjan Singh said he was posting Idris to command No.2 Wing at Pune or No.4 Wing at Agra, whichever he preferred. Both were important operational wings, and the command of both of which would be soon upgraded to the rank of Air Commodore. Idris had never, ever been given a choice of postings, and was deeply touched by the Air Chief's gesture. He chose Pune which had a couple of fighter squadrons, a Canberra bomber squadron, plus a maritime reconnaissance squadron

with Super Constellation aircraft. From our personal point of view, Pune was preferable, as both our widowed and ageing mothers would be more easily accessible.

The Air Chief had mentioned to Idris that until the wings were formally upgraded, he would be posted at Udhampur as Air Officer Commanding Jammu and Kashmir. Idris' delight and excitement knew no bounds: "Imagine", he told me, "I am getting the command of a vital sector of the country which, for all these years I have had very few opportunities to visit." But he added that, very sadly, since in Udhampur there was no married accommodation, he would have to leave the children and me behind. It would, however, be for only 2-3 months and during that period, the children and I would be permitted to continue to stay at Raksha Bhavan, where we had spent almost a year whilst Idris was at the National Defence College. A smaller apartment was then allotted to us and the children and I moved there. It was good to be able to stay on in such a safe and secure place.

While we remained in Delhi, Asad and Asgar were able to continue to study at the same school. Mariam, however, who had already been to eight different schools because of our numerous Service transfers, had asked us if we would get her into the Welham Girls High School in Dehra Dun, so that she could be in one school for her two final years. She was, fortunately, able to get a place at Welham and finished her two years of schooling there. Two years later, we were thrilled when her final results were outstanding, with her coming 7th in the whole of India, with the highest aggregate of marks that any girl had ever got at Welham until then. Asad and Asgar had also been doing very well at St. Xavier's, so we decided to keep Asad there if possible, as he was also in the final three years. Asad stayed on as he was accepted as a boarder. He remained there till the end of his schooling, and did very well in the final exams, finishing his schooling when he was just 15 years old. Fortunately, he became 16 by the time the college year started. Asad was always especially good at mathematics and also at games, and captained the badminton team; he was also on the school hockey and swimming teams. For both of them, living in boarding houses at their schools and managing on their own was something we felt they were happy to try out as the experience would help them to become more independent, whilst still in a secure environment. Asgar was young enough to come to Pune with us. He was, therefore, the only one of our children who actually lived in an Air Force camp.

CHAPTER 16

In Command of 2 Wing, Pune

The months passed quickly, and soon Idris was back in Delhi from Udhampur. We packed and decided to drive to Pune in the wonderfully reliable Austin Cambridge Idris had acquired when we were in Washington. It was a wrench for us not to have Mariam and Asad with us, but we were lucky that we could have Asgar with us. So, ultimately, early one fine morning, just the three of us — Idris, Asgar and I — set off for Pune, with Idris quite excited with the thought of driving all the way. We took about three days so that we wouldn't have to race against time and could make it a fun trip. On arrival, we initially stayed at the MES (Military Inspection Services) Bungalow, which was wonderfully relaxing.

A day later, Idris took over command of No. 2 Wing. His joy and excitement had to be seen to be believed. "Just imagine," he told us, "I have the command of a major operational wing of the Air Force." The icing on the cake was that it was a base for two fighter squadrons, both soon to be reequipped with Maruts, which Idris hadn't flown before. There was also a Canberra bomber squadron and a maritime reconnaissance squadron equipped with the huge four-engine Super Constellation aircraft, all of which he would now be able to convert onto and fly. He told me with a smile, that as the Commanding Officer, he could now do so without having to get clearances from a higher formation! Idris later converted onto each of the aircraft and flew them. He said to me, "In any case, on a more serious note, I strongly feel that if one is in command, on principle, one should have at least first-hand knowledge of the capabilities and problems of each, for any meaningful discussions with the Commanding Officer and his team of officers and men. If I remember the

Canberra bomber: #35 Squadron

Marut fighter : # 220 Squadron

Super Constellation: Maritime Recce Aircraft: # 6 Squadron

sequence right, first it was the Canberra, then the Maruts and, finally, even the large four-engined Super Constellations."

We soon moved to the house allotted for the Air Officer Commanding (AOC) the station, which was a very comfortable and well-planned three-bedroom house. Along with some our own furniture, it became very adequately furnished. There was a garden and an enormous compound providing the entire privacy one could want. Idris continued as Air Officer Commanding 2 Wing at Pune till August 1970; it was, as Idris often put it, amongst the most stimulating, satisfying and emotionally rewarding assignments of his life in his beloved Air Force. With all the officers and men, including the civilians, and the families, the total population was perhaps around ten thousand. They were really a fine group of people and for Asgar also, it was like being with a big family — of around ten thousand! Activities and games were shared; and welfare work was results-oriented with immediate improvement possible in the lives of those who had problems, or for other needs of the station.

For those who did their flying very early in the mornings, a good breakfast for the crew was then organised. This had been strongly advised by the medical authorities. Idris was able to establish a canteen adjacent to the crew rooms, where those who flew early in the mornings, and were often unable to have a proper breakfast before flying, were thereafter provided freshly-cooked, nourishing food eaten comfortably before they commenced their flying. In this context, I must mention that several years later, before retiring as the Air Chief, free rations for the air crew were proposed by Air Headquarters and approved by the government.

Saving a Damaged Aircraft

Idris was happy to be able to get plenty of flying and, on one occasion, I learnt from the squadron officers, that he was able to avert a serious accident, which, in fact, saved the aircraft. One night, Idris had a sortie on a Canberra tasked for live bombing practice with 1,000-lb. bombs. After dusk, soon after taking off, Flying Control called him on the radio transmitter to say that parts of the aircraft tyres had probably come off and to properly assess the damage, it would be advisable if he could fly past the control tower with his undercarriage down, to enable them to see if there was any serious damage visible. Idris, therefore, instead of proceeding to the bombing range, flew past the tower and then circled the airfield to await a further call. They soon informed

him that clearly the rubber retread of one of the tyres on his aircraft had come off and that a normal landing would be dangerous, in fact, not possible, as the plane may well swerve off the runway and crash. He was advised to carry out an emergency landing as soon as possible.

However, this could not be done immediately, Idris later explained to me, as he had a full load of bombs and of fuel, with the result that the all-out weight of the aircraft was far above the maximum landing weight. He, therefore, to start with, flew to the Ahmadnagar live bombing range to enable him to first drop his bomb-load. He then returned to the airfield but continued flying to burn up as much of the fuel load as possible, to reduce the landing weight to within the permissible limit. Retraction of the undercarriage was also not advised, as when retracted, some piece might foul up the system, and then it may not have been possible to lower the undercarriage at all. He, therefore, circled near the airfield with partial flaps down and with a high throttle setting, and in a little over 45 minutes he had burnt up enough fuel to permit an emergency landing.

By then, of course, it had become quite dark. The navigator in a Canberra, Idris told me, normally sits in the nose of the aircraft, which under the circumstances was particularly risky. Flight Lieutenant Chandrashekhar, his navigator on that flight, Idris said, was one of the finest he had flown with, and, to his credit, he remained completely composed. Carefully nursing the engine, Idris tried to do as smooth a landing as possible, knowing that the rubber-less tyre which was badly damaged on one side, could burst on the touchdown, and may well cause the aircraft to swing off the runway, completely out of control. Happily, Idris told me after he got home, that it was perhaps the best landing that he ever made on the Canberra! He added that with very gentle use of the brakes, he was able to bring the aircraft to a complete stop. Both he and Chandrashekhar must have heaved sighs of relief; not only was a Canberra saved, but, the two of them had received a new lease of life. Idris was happy when he received a commendation from Command Headquarters. The officers on the station were, however, most upset as they were expecting him to get a decoration (!) — some even protesting when it was not announced! However, Idris and I were deeply grateful to God that it had ended so well, and both Idris and the aircraft were up and flying again the very next morning!

The Chief's Visit

In 1970, Air Chief Marshal Lal visited Pune when Idris was still in command of the wing. After a very comprehensive inspection, the Air Chief apparently complimented every Squadron Commander and also, very particularly praised the organisation and upkeep of the Wing Maintenance hangar, under Wing Commander Gursharan Singh, and his deputy Squadron Leader Shashi Ramdas, both of whom Idris was very fond of and greatly admired. On returning to Idris' office, after the normal briefing on the operational, maintenance and administration status of the wing to the Air Chief, he again complimented Idris on the maintenance system that had been evolved at the wing. Idris, of course, again stressed that credit for the concept and the standards achieved was due to the outstanding technical team of officers and men he was fortunate to have.

Idris receiving the Air Chief, Air Marshal P.C. Lal at 2 Wing, Pune

After completing his briefing on the wing, Idris apparently requested the Chief if he could have a few minutes to share, in complete confidence, some thoughts with him, a request that the Chief agreed to without the slightest hesitation. When they were entirely alone, Idris said that since another armed conflict with Pakistan was clearly

imminent, he would like to present a couple of ideas for the Chief's consideration. The Chief was clearly delighted. Many, many years later, in fact only after retirement, Idris told me that after first underlining to the Chief the very great importance of incorporating the element of surprise, which, he said, in his view, had seldom been given the importance it deserves, it was in this context, Idris went on to say, that he had a suggestion or two for the Air Chief's consideration.

To help achieve air superiority, Idris apparently went on to say, even for limited periods, bombing key enemy airfields, including runways, was well-known. But most key airfields nowadays had fairly effective arrangements for quick and, at least temporary, repairs. It is in this specific context, he continued, that he had two suggestions to offer which he felt, if incorporated in our offensive air operational plans, could well exploit the principle of surprise. To this day, I don't know what Idris' suggestions were. But all that I do know is that when Idris came home after seeing off the Chief, he mentioned that the Chief had apparently gone out of his way to compliment Idris not only on the operational preparedness of 2 Wing, but very particularly stated that he was really delighted that Idris had been applying his mind, and in such a concrete and constructive way, to the likely conflict ahead.

The atmosphere at 2 Wing, Pune, had become one of mutual sharing and caring. This extended to the villagers of the neighbouring Lohegaon village, who desperately needed water with which they were, on Idris' directions, helped by the Air Force whenever the shortage was very acutely felt by the villagers. We both admired the Sarpanch and villagers a great deal when they undertook and implemented, entirely on their own, a lift irrigation project which brought water for their farms, from quite a distance. Both Idris and I interacted with the people of the village and under Idris' direction, the station helped to the extent possible, but always ensuring, on Idris' orders, to the extent that each initiative was within rules and regulations. For example, in getting long overdue compensation to the villagers for the temporary use (with their consent, of course) of some fields for short periods for air armament practice. Of course, this was done not with bombs, but practice rockets and gunnery, and even these only with practice ammunition.

A great deal was also done to build up and improve the station facilities. A major achievement of which Idris — along with the rest of the wing — was particularly

proud, was the construction and inauguration of an open-air theatre. For the planned open-air theatre, Idris, with a selected team, organised a daily *shramdaan*, on working days. For three months, instead of the very regular afternoon games/physical training, he, with other officers and airmen, plus the civilians, gave a hand for an hour or so on most afternoons. They handled the large quantities of stone and other material that was required more or less on a continuous basis for the construction work. The technical support was most generously extended by the EME, the Military Engineering formation in Pune. When completed, Idris invited the (then) Air Chief Marshal Arjan Singh to inaugurate the facility, an invitation which he immediately accepted. Idris and everyone at 2 Wing, families included, strongly felt that the theatre should, on inauguration, be named the Arjan Theatre, and so, with his consent, it was. Very soon it became very popular, particularly as the existing indoor cinema hall of the station was now inadequate as the number of people there, including families, totalled around ten thousand. In fact, it was an important, perhaps compelling reason for Idris to take the initiative in the first place. In the evenings, with regular cinema shows and not infrequent variety shows organised by the talented personnel of the station and their families, it very soon became the most popular meeting ground for all the families, especially as the Air Force Station was far from town. The generally salubrious climate of Pune also helped greatly. Having constructed it themselves, it not only gave a sense of satisfaction but also of great pride. Idris was delighted when recently the Commanding Officer told him how very popular and intensively utilised the facility still is.

Air Chief Marshal Arjan Singh inaugurating the Arjan Theatre built through '*shramdaan*' at 2 Wing, Pune

Other wonderful memories are of times when the officers and their wives would, usually after a late squadron party, suddenly land up at our home any time between midnight and 2 am (!), and keep honking their car horns at the gate till we woke up and called them in. It became, in fact, great fun, and we served up whatever eats and drinks were available in the house. It was most informal and enjoyable, and there was plenty of sound and laughter.

Another memory comes back to me: once a very senior officer was expected to visit the station. There was no VIP room, so a room in an annexe to the Mess was to be done up. It had a single bed, a writing table, a chair and a cupboard. Idris sanctioned five hundred rupees to do up the room for the visit, clarifying that the amount was all that the station's non-public funds could afford for this purpose. Knowing how tight such funds were in those days, I first bought something for the floor, and my budget permitted me just a plain 6-foot by 3-foot cotton carpet, but, fortunately, I found one in an attractive colour. Another two hundred rupees got the inexpensive curtains for the windows, with a bedcover to match. Some hangers for the cupboard were added, as well as soap and a soap dish, plus fairly simple bed and bath linen. The ashtray and a painting were donated by us! With the addition of a few potted plants from the Mess garden and a beautifully arranged vase of fresh flowers on the table, No.2 Wing welcomed the incoming guest. It is hard to believe these days, but in earlier years, it was really a struggle to provide funds. Looking back, it was a challenge for me who, after all had had the good fortune to be trained as an interior designer in the US. Soon thereafter, I was requested by the President, Mess Committee (PMC) of our Mess to do up the bar at the Officers' Mess also. Here again, funding was only slightly — very slightly — more generous! However, it was a challenge and quite good fun with various innovations that turned out well.

Idris had been keen on photography since he was seven years old, and wherever we went, his movie camera was often with him and he made lovely films. We, therefore, have a great collection and record of wonderful visits, of beautiful places and also of the activities that we shared with the three children. These include not only picnics but also amateur dramatics as they performed different plays, apart from Rumplestilskin in Shillong. Whilst in Pune, they acted and produced Hansel and Gretel, both mostly directed by Mariam. Asgar was the most reluctant to participate, as his great concern was that his friends on the station might spot him acting in some odd role. However,

he did participate which was very sporting, because at the time he also had his arm in plaster for a broken forearm. Asgar was a very keen swimmer and a bit of a swimming champion too.

As Mariam and Asad were usually away at school as boarders during their last years of schooling, these periods of togetherness in their holidays were very precious for all five of us. Towards the end of our stay at 2 Wing, when it was time for Idris to be posted elsewhere, Asgar was put into the Air Force School at Palam as a boarder too. We could have been posted anywhere, and every time we got transferred, it seemed to be towards the end of a school year. When we arrived at Pune too, I remember I had to go from one school to another to get Asgar admitted. The principal of one school told me, "You Defence Services people... you always come in the third term and if we take you, it disturbs and disrupts our class at a crucial time. Please go elsewhere." In most postings where we arrived well into the third term of a school year, when the children had finally been accepted, they had to really study non-stop. They were often up till late at night and again early in the morning to catch up with others who had studied there through most of the year. They had to really work endlessly to enable them to face the final examination within a month. There was no time for extracurricular activities either. Very much later, this problem was eased somewhat for future postings for others, as the rules were improved so that when the husband was transferred elsewhere, the Air Force allowed wives and children to retain their official accommodation till the children had finished the year's final examinations in the same school. In our children's case, this facility was lacking.

A very popular project at the Air Force Station became the Best Home and Garden competition for married NCOs and airmen; as it had been when Idris had first started it at Air Force Station Hyderabad in 1959, the very first station that he commanded. It was a success in not only improving the appearance of the residential areas, but also in establishing a cordial relationship, particularly amongst the officers' wives and the families of the men, contributing greatly to the general morale of the station. There was a Ladies Club for the wives of non-commissioned officers and other ranks as well as an Officers Wives Club, both full of wonderfully active women. I got to know the senior NCOs' wives well too, and they were especially helpful, active and enterprising.

There were also sports in the afternoons, where Idris joined in games like hockey with the others on the station. Idris, Asgar and I were also able to join in sports with the families of the other officers too. There were badminton matches for us and the children played cricket or went swimming, as all these facilities were available. My days were busy with the usual numerous activities as well as efforts to create awareness of the need for small families, and for pre-primary schooling for children of Class 1V civilians. There was a great paucity of funds, and utmost economy was the norm in those days. Idris offered us the use of an old, disused and rather battered Nissen hut. It was most welcome, and after fixing it onto a base of bricks and cement with the help of the Class IV personnel and their ladies, it was cleaned and spruced up. It became their children's KG school. There were some wives of the Class IV personnel with severe health problems who needed humane attention, which was arranged. Family planning was also an important requirement which I took up along with the health project.

After some time, my interaction with the Class IV personnel brought about a strangely gratifying result. One day, a few senior representatives from the union of defence workers in Pune came to tell me that the union members of all the defence establishments were going on strike. But he added that in view of the care and concern extended towards the workers and their families at this Air Force Station, the union members had decided to refrain from going on strike at our station. He added that, therefore, their personnel at our station would be the only ones not required to do so. It was so very heartwarming and gratifying. Strangely, many years later in 1992, when I was made the Chairperson of the unique and special Bal Bhavan established by Pandit Nehru in Delhi, there was some union problem going on there. I was surprised and very touched when some of the senior persons of the All India Labour Movement or union came all the way to meet me in Hyderabad. They briefed me, and said that they wanted to come and do so as they had known me to be caring and sympathetic to the workers since our Pune days! I was never quite sure of which these unions actually were, so I may be mixing them up!

My first exposure to AFFWA came about in 1971 while Idris was still in Pune. One day, a letter for the Air Officer Commanding 2 Wing arrived with a query in it for me. It was about a proposal by Mrs. Ella Lal, the wife of the Chief of the Air Staff, for establishing an Air Force Wives Welfare Association. I was asked to send back my comments. I now find it quite amusing that my reaction and reply, both, were negative! I wrote back to say that we already had Ladies Clubs with ladies actively involved

in contributing and working hard for whatever they could do for station activities and welfare, and with funds being collected for the Indian Air Force Benevolent Association. I added that this was so in most Air Force Stations all over the country, and that collecting funds for the station welfare work entailed a great deal of effort. I added that, therefore, one more organisation would only create a duplication of effort, and of accounting, and need not be implemented!

Soon after this, Idris was transferred to Delhi where Mrs. Lal informed me with a mischievous smile that the new Air Force Wives Welfare Association, AFWWA, had been established and that I had been elected in absentia as the Honorary Secretary of the organisation! Mrs. Lal never held it against me that I had not gone along with her plan originally, and, in fact, she joked about it. She has always been a very special and inspiring lady, with the finest qualities of integrity and values. She was also most caring and kind, gentle but very firm about doing what was right. I have looked up to her with great admiration and respect as well as affection as for an elder sister.

While we were at Pune, as a trained artist, I undertook a very interesting art assignment. I had seen an advertisement in the newspapers for a senior artist by Air India. Having earlier been trained in Fine Art, Miniature Painting and Modern Art, I applied. I was asked to come for an interview and bring along some of my work. I was unable to accept the job as it entailed being in Bombay every day during office hours, and occasionally even longer. However, I was then asked to do a mural for the Air India pavilion being put up at EXPO 70 in Osaka, Japan. It was for a 24-foot-long mural which would be on a curved wall which was six-feet-high. Part of it would extend to cover all four sides of an 8-foot pillar. They had liked my sample of a painting which I had done of abstract birds with peacocks all over on a rough 3D finish. This fit with their theme of that year. I took this on, and got to work on the 24-foot curved boards and then the painting in relief in the style which I had studied with a great Italian artist, Pietro Lazare, in Washington D.C. It was done on several curved panels to fit the curve of the wall on which it was to be fitted. Experts came from Air India to examine it, and after it was approved, I was asked to fly to Osaka to have it installed on its curved wall. I felt that it would not be right for me to leave my home and other responsibilities, so I said that I was unable to go. An expert was then sent to see to the panels being carefully packed and sent to Osaka, where it was displayed on the 24-foot curved wall at the Air India pavilion of the EXPO 70 World's Fair. Two

years later, in Delhi, I again responded to an ad in the newspaper for a mural for a government building in an open competition. Mine was selected out of all of them. I was totally overjoyed. I remember requesting that wonderful couple Raj and Ramesh Thapar to be kind enough to come and see it critically, which they were kind enough to do before I had completed it.

Our years in Pune passed far too quickly, and when Idris was transferred to Delhi even though it was on promotion to the rank of Air Vice Marshal, he was really very saddened when he left. Till today, he often says that, apart from his three years as Chief much later, for him his years in command of No.2 Wing were the most challenging, the most stimulating and the most heartwarming of his entire career. Every officer of the flying branch, he adds, yearns for such an appointment but very few, of course, are as fortunate as he was. Whilst handing over command to Dilbag Singh, his very able successor, Idris addressed the 2 Wing personnel with a full heart brimming with emotions, but with satisfaction also that all was well with 2 Wing and that its future under Dilbag would be secure. The following is his farewell speech on handing over command:

It is hard to believe that more than two years have passed since the station held a similar parade at which I took over from Group Captain S. Singh. It is now my turn to step aside and make way for Air Commodore Dilbag Singh, a fine officer of high repute and an outstanding record of service.

First of all, may I compliment you on the standard of the parade? It is obvious that you have made a very special effort not only in ensuring a very smart turnout, but a very high standard of the movements. I feel so proud of your performance.

At a moment like this, when one's thoughts are tinged with the sadness of parting, but one's heart is also filled with pride, I find it hard to put into words my very deep sense of appreciation of the effort that all of you have made, both Servicemen and civilians alike, in the performance of your duties. Your achievements have, truly, been of a calibre that any station could be proud of; achievements that have certainly enabled me to hold my head high when any one has visited us — from the Chief of the Air Staff downwards.

I want you all to know that, very recently, I put up a board in my office, and

inscribed on it the names of all those who have been decorated or commended for Meritorious Service, either by the CAS or the Air Officer Commanding in Chief (AOC-in-C). I cannot tell you how proud and happy I was to note that in these last two years alone, five of our station personnel have been decorated, and no less than 21 Commended by the Chief of Air Staff or AOC-in-C. It is a record of which any station would be legitimately proud.

Throughout my time here at No. 2 Wing, I have watched with great admiration the manner in which 6 Squadron has wrestled with the problem of low serviceability, managing always, somehow, by working night and day, to meet all its major commitments.

Idris and then Air Commodore Dilbag Singh
at 2 Wing, Pune

I have watched with great admiration also, the two Marut squadrons, the skills and great courage of the pilots and the determination and dedication of the ground crew, to do their very best to make this, the first fighter aircraft designed and produced in our country, a more effective weapon system. I have watched with admiration our Canberras, No. 35 Squadron, never ever failing to achieve its monthly flying task, coupled with their truly outstanding performance, in every single exercise or weapons meet that they have participated in. Well done, our bomber boys. I have watched, with deep satisfaction, the quiet but exceptionally competent manner in which all our Missile Squadrons have conducted their training programmes; always, without fail, earning high praise from all who have come to inspect them.

And, finally, I have watched with a feeling of great happiness and pride and admiration, the manner in which all the sections of the headquarters of this wing, whether they be in the hangars or sitting behind desks in offices, how all of you, each and every one of you, have worked, and worked with single-minded devotion, as one splendid team. Day after day, month after month, year after year, you have ensured that every unit with us and the wing as a whole accomplishes its tasks in the best traditions of the Air Force. It has truly been an unforgettable experience. By your loyalty, your devotion to duty and your competence, you have made this assignment of mine, perhaps the most cherished in all my years of service. For all that you have done for our wing, our station, I thank you. I now leave you in the competent — and caring — hands of your new Commander, Air Commodore Dilbag Singh. God bless you all.

CHAPTER 17

Shaping the Future of the Air Force

Air Chief Marshal (ACM) Lal had time on his side, but the manner in which he prepared the Air Force for the 1971 War was truly admirable. ACM Lal's farsightedness, Idris always said, did much for the efficient reorganisation of the Air Force, particularly of Air Headquarters. For the Air Force's operational responsibilities also, as demonstrated, Idris said, by the conduct of the 1971 War, the Air Chief had splendidly planned and organised every aspect of the strategy as well as optimum decentralisation of command and control of all the formations. Air Chief Marshal Lal soon transferred Idris to Delhi where he, with his remarkable foresight and imagination, had obtained the government's approval for the creation of the new department for a well-balanced growth of our Air Force within the confines of the financial resources made available from time to time. For this, he appointed Idris as the first Assistant Chief of the Air Staff, Plans [ACAS (Plans)].

This reorganisation was part of one of ACM Lal's farsighted initiatives for our Air Force. Idris had, under him, the DASR (Directorate of Air Staff Requirements) for Weapon Systems and DPP (Directorate of Plans & Programmes – organisation, manpower, etc); the DPG (Directorate of Project Groups) was also with Idris primarily to oversee modifications/upgrading tasks entrusted to industry (HAL mainly). Air Chief Marshal Lal, in his very comprehensive briefing to Idris on arrival, specially stressed that part of his important direct responsibilities would be to ensure that all phase-out/phase-in plans were always dovetailed with, and fully reflected, the financial resources likely to be made available by the government. It was primarily for this reason, he emphasised, that the newly created Directorate of Financial

Planning had also been placed under Idris as ACAS (Plans). This, he said, had to be an *essential* ingredient of all planning. And rightly so, Idris once mentioned to me also, as resources were really tight in those days — far, far, far from the overflowing coffers of recent years!

Meanwhile, in August 1970, whilst Idris was still in Pune, Air Chief Marshal Lal took up Idris' suggested measures for disrupting the use of enemy runways in the event of another war. On his arrival in Delhi, the Air Chief called Idris and told him that the task of implementing his proposals had been entrusted to the Air Officer Maintenance, Air Marshal Malse, and he asked Idris to check the status. Eventually, as Idris mentioned to me very much later, that it was, in a limited way, used in the 1971 conflict against Pakistan and achieved encouraging success. This was confirmed subsequently in a way, when senior Air Force officers of a country in West Asia as well as someone in London raised the subject with Idris, seeking more information, and questioned Idris on some novel device that our Air Force had so successfully, though admittedly in a limited way, used in the 1971 conflict with Pakistan. Each also asked if Idris could enlighten them. Idris, of course, as expected, expressed surprise and feigned ignorance. Deep down, of course, we were quite thrilled.

On the evening that Pakistan's aircraft attacked our airfields in 1971, Idris and I were driving down Lodi Road outside Khan Market in Delhi at dusk, when we were stopped by a group of people. They told us and all other drivers of other cars driving by, to switch off our lights, as Pakistan had attacked our forward airfields on the Western front. Mr. Bahri himself (of Bahri Sons bookshop at Khan Market), and several other shop owners were, I noticed, amongst them on the road. Idris went straight to Air Headquarters while I went back to our home. The counter-air operations by our Air Force included, Idris soon learnt, the use of the novel weapons suggested by Idris to temporarily immobilise carefully selected enemy airfields. In the days that followed, our Air Force also most successfully attacked the oil storage facilities at Karachi (see the news article, "IAF's War Hunt in Karachi" at the end of this chapter).

Idris greatly respected Air Chief Marshal Lal, but felt free enough to suggest to him that for the duration of the war, 'Plans' could be set aside, and he (Idris) be given an operational assignment. Idris tells me that, with a smile, Air Chief Marshal Lal said, "I tell you what, take 2-3 officers of your choice and an AVRO, and visit the front

line units/formations and be my personal liaison officer, and concurrently, personally evaluate the strength and limitations at the front line in each sector." The Chief added, "And, what is very important, for the long term, it will help you make a realistic assessment of our organisation at important levels and the strength/limitations of our weapon systems. This first-hand knowledge will help you, after the war, in your long-term planning of our Air Force." What a Chief!!!

Additionally, the Chief had added that, as the need arose, he wanted Idris to be available as his special representative to receive, from Idris, should the need arise, emergency feedback. He concluded his briefing by adding, "Latif, feel free to call me directly if you find that a very urgent operational need warrants it; I may also similarly use you to convey critically urgent and important instructions, should *I* find the need." Incidentally, it was in this context that, when the Circuit House had to be attacked within hours, he called Idris who was at Eastern Air Command in Shillong. It was early at eight in the morning that the Chief's call to Idris came, stressing that it was very important that the strike be timed exactly at 11 am. Idris was instructed by the Chief not only to brief the C-in-C and SASO (Senior Air Staff Officer) but also, personally, to monitor its planning and implementation.

After the Chief's briefing on his responsibilities during the conflict, Idris, to start with, flew the very next morning with his small team, initially to Shillong the Headquarters of Eastern Air Command from where the critical battles in then Pakistan's eastern wing were being fought. He spent all of the next day in detailed briefings by, and discussions with, the C-in-C, Air Marshal Dewan, and his deputy Air Commodore Devasher. Then, taking advantage of night flying, he paid quick visits to the Northern and Western Sectors, spending a day or two, as needed, with key formations and stations. Idris, with his team, then flew back to Shillong for the concluding days of the war. It was soon thereafter that Idris (as briefly mentioned earlier), received the call from the Air Chief at eight one morning, saying that they had reliable information in Delhi that Dr. Malik, the Governor of what was then East Pakistan, had convened an emergency meeting to consider the military crisis that their armed forces were facing as a consequence of our air and land offensives. The meeting was scheduled to be held at 11 am at the Circuit House at Dhaka and he (Air Chief Marshal Lal), wanted an air attack to be mounted on that building exactly at that time. Idris conveyed this to Air Marshal Dewan who was the Commander-

in-Chief of our Eastern Air Command. The latter said that Idris should liaise with Air Vice Marshal Devasher, the SASO and also personally monitor and help, where necessary, to plan and oversee the strike operation.

Unfortunately, no one knew exactly where the Circuit House in Dhaka was located. A frantic search commenced for a correct map. Finally, a civil map was located but even this showed two Circuit Houses! Telephonic enquiries and frantic calls were made to Calcutta to get more information. It was already nearly 10 am. Finally, quite reliable information was obtained but they were warned to ensure extreme accuracy as the residences of foreign diplomats were also in the vicinity of this Circuit House. This was conveyed to the field commanders.

As suggested by Idris, Group Captain Mally Wollen, an exceptionally competent officer, who had been positioned at Gauhati, was asked not only to personally oversee the selection of the best crews, but also to directly monitor the entire strike operation. Eight fighters armed with rockets were launched. At ten minutes past eleven, the returning aircraft radioed to the Operation Room that had been set up at Headquarters Eastern Air Command, "Mission accomplished." Later, one of the UN representatives in Dhaka mentioned to some of our diplomats that Dr. Malik had just convened the meeting which they thought was at a secret location, when our IAF aircraft struck. Photographs later revealed that every rocket had hit the target, with not a single one going astray. The effect on those at the conference was apparently so devastating; that Dr. Malik told all those assembled there that the time to surrender had come. He is said to have spread out his prayer carpet thereafter to say his prayers. The surrender ceremony is well documented; however, we were informed about this by those who were present. Air Marshal Dewan the C-in-C of Eastern Air Command, and Group Captain Chandan Singh had both flown in to attend the ceremony. Air Marshal Dewan is said to have asked General Niazi what made him decide to surrender. The General was reported to have walked up and pointed to the wings on the chest of Chandan Singh's uniform and stated, "This (the Air Force) left us no alternative."

Soon after the surrender of the Army in East Pakistan, Idris was again able to fly to our different frontline Air Force stations on the Western Sector to make a final assessment of the capabilities and shortcomings of our frontline formations for himself. He was also, at the end of each day, able to give the Air Chief, whenever

called for, an up-to-the-hour and more refined and comprehensive summary of his assessment from the area at which he was.

After the 1971 War, Idris as ACAS (Plans) prepared his report as an assessment of his observations of aircraft and infrastructure performances in different sectors as 'pointers' for future planning for the Air Force. In this connection, it is worth mentioning that, en route to Shillong, his very first visit was to the Gnat squadron commanded by (then) Wing Commander Sikand in the Eastern Sector, where he had an excellent opportunity to analyse the strengths and shortcomings of the Gnat. He was particularly impressed by the almost unanimous view that, subject to a few improvements to the control system and its comparatively very limited range, it had a definite future, keeping in mind, specially, the extremely tight budgets that the government was very firm it could not augment. Idris once, much later — in fact, after his retirement, confided to me that, after extensive discussions with senior, very experienced pilots such as Mally Wollen, he was convinced that the Gnat should be optimised. Basically, the Gnat's excellent performance in the 1971 War, coupled with its low cost, convinced Idris and his specialist staff, that with an improved control system and radius of action, it deserved to be given a longer life. This was particularly so, keeping in mind the extreme budgetary constraints within which the Air Force had been compelled to work. The Ajeet was the outcome.

But the Ajeet, Idris was fully aware, would, at best, be a fighter with very limited ordnance. Coupled with the marked obsolescence and consequent limitations of the Canberra, it was of crucial importance that our strike element should have top priority in our modernisation programme. To begin with, therefore, he had very extensive discussions with the two PSOs (Principal Staff Officers) directly involved, namely, the Deputy Chief of the Air Staff (DCAS) and the Vice Chief of the Air Staff, (VCAS), and having served under both, he knew them well. Both, fortunately, fully agreed that augmenting our strike capability should receive top priority. But, whilst one very strongly favoured a multi-role fighter for air defence and reasonable strike capability, the other, equally strongly, favoured a pure bomber!

Fortunately, a few days later, the Chief, Air Chief Marshal Lal, whose confidence he also enjoyed, called Idris to his office. Before leaving, however, Idris decided to seek his guidance on the problem he faced. With a charming smile, the Chief indicated to Idris

that he fully understood his predicament. He then asked him for his own view. Idris was very clear in his mind and said, "Sir, we all know that no war can be won without meaningful offensive strikes. We still have some comparatively modern fighters, but the 1965 and 1971 Wars have conclusively demonstrated the severe limitations of our only bomber, the Canberra. There is no doubt in my mind that what our Air Force badly — very badly — needs is a bomber; not one of the 'biggies' of World War II but an aircraft that can carry a sizeable bomb load, say around 8,000-10,000 lb. coupled with really up-to-date, low-level navigation and precision bombing avionics." The Chief's response was, "I agree with you," but he added that, unfortunately, especially after World War II, the word 'bomber' conjured up, amongst many in government, images of callous and large-scale destruction of life and property. He added, "I fear that any mention of the word 'bomber' will invite strong opposition! Can you think of another name?" The next day, Idris saw the Chief again and said, "Sir, why not call it a 'Deep Penetration Strike Aircraft' — to add a spin, the DPSA, in short?" "Not a bad idea, let's try it," was the Chief's response — to Idris' great delight — and relief.

When the Chief then asked, "Have you any particular aircraft in mind?" Idris replied, "Sir, I've already had several discussions with DASR (Director of Air Staff Requirements) and one or two very experienced bomber pilots who strongly underlined the great importance of twin engines, which I also feel is very important for survival over long distances at low levels, whilst both the other possible contenders, the Mirage F1 and the Swedish Viggen, were single engined." Furthermore, at low levels, so critical to escape detection by enemy radar, the low bypass engines of the Jaguar gave it a much better radius of action, he observed. And, finally, and importantly, the Jaguar, unlike the other two, had a tailored low level navigation and attack system. The Anglo-French Jaguar, therefore, in several respects, had a definite edge, especially for the role required. "If you agree, Sir, we'll put it up for scrutiny by the concerned PSOs and then up to you for your consideration." To cut a long story short, the Jaguar was unanimously accepted by the Air Force. Though the government also accepted it in principle, but, primarily (presumably) because of the very high cost involved (if I remember right, Idris mentioned to me once, that each was in the region of a little over a crore) the final 'go-ahead' was, because of paucity of funds, deferred year after year, after year; Chief after Chief after Chief! Finally, almost eight years later, it landed in Idris' lap, when in 1978, he took over as Chief! But there is more about it in a later chapter.

Even at that time, Idris strongly felt the need also for a light combat battlefield support fighter aircraft. This was not for air defence but with acceptable escape capability, an indigenous, but relatively less sophisticated and, thus, a relatively inexpensive aircraft, as the government's purse was, Idris often said, very, very tight then. In fact, funds were so short that sophisticated aircraft were too difficult to procure with what was available. The limited capability of HAL to design and develop aircraft was another complicating/restricting factor. While the MiGs could fulfill these tasks to some extent, they had already been in service for several decades and would need replacement at the latest from the early 1990s, substantially modified to keep it within the available resources. His rationale, he confided to me, was that in times of war there was always the need for substantial numbers over the battlefield areas in support of the Army. The relatively easy tasks such as tanks, artillery positions, etc. could, however, be satisfactorily engaged with the relatively small, and unsophisticated Gnat/ Ajeet, rather than risking the more expensive aircraft which had been employed in earlier conflicts. If indigenous, and, therefore, available to us in larger numbers, it may also permit saturation of the battlefield area in our favour. Indigenisation, Idris had always felt, was most desirable as it was high time that our country at least started moving away from total reliance on imports.

When submitting proposals to the government, whether it be for manpower or aircraft, the government rightly asked that each proposal should be presented as part of our long-term plans. Idris recounts, "I decided to have two components: a Five-Year (rolling) Plan, and a 20-year Perspective Plan. The Chief (and the government) approved. To prepare these, I asked the Ministry to indicate, at least in broad terms, the financial resources on which our planning of induction of new aircraft could be based. They were most reluctant; but ultimately, recognised that without it, no plan would be meaningful. After a series of high level discussions, we were told that we could base it on a budget that, for modernisation of our squadrons, could count on an annual cost escalation of 5 percent"! Idris adds, "For months I argued (at every level I could reach!) that whilst for the Army which is a manpower-intensive Service, cost escalation may not be more that 1-2 percent or so a year; but for the Air Force (and the Navy) where expenditure is largely on equipment and, sadly, largely dependent on international suppliers, it was, as was well known, over 10 percent. If their directions were to be implemented, we argued, our force level, already below what had been approved earlier

in principle by the government, would clearly go down even further. Whilst they agreed that this was not the government's intention, for purposes of planning, 5 percent would have to suffice. Thus, unrealistic though it was, Air Headquarters, at the highest level, also accepted that for planning purposes at least, it would be wrong to ignore the government's constraints. And these, we all also knew, were very real."

Though Idris had prepared a Five-Year Plan and a Perspective Plan for 20 years, after the 1965 and 1971 Wars, a review was not only desirable but urgent. This would have to be a review that would cover not only our weapon systems but also all essential facilities to support their operations, which again, not only had to be strengthened, but updated and modernised. Taking into account the limited radius of action of our fighter strike force at the time, Idris, therefore, added Advanced Bases (then named FBSUs — Forward Base Support Units) by which targets across the border would come within strike distance. Keeping cost considerations in mind, three such Forward Base Support Units to start with, to cover the Rajasthan region and the north were decided upon, and given the basic infrastructure for fighter operations including items like POL (Petrol, Oil, Lubricants) stores, blast pens and maintenance facilities. The requisite minimum manpower was also included. Once Idris had finalised a firm Five-Year Plan and a 20-Year Perspective Plan for the Air Force, they had been sent to the Defence Secretary Mr. Sushital Banerji, who was an outstanding civil servant. Some time later, when Idris went to see him on some other matter, he asked Sushital what he felt about the plans which he (Idris) had made. To Idris' surprise (and disappointment) Sushital knew nothing about them. On Sushital's stern enquiry, it was found that though his staff had received them, they had not put the plans up to him. The papers had just been parked in a file cupboard. Sushital, being the gentleman that he was, immediately apologised, and said he would go through them on a priority basis right away.

Early in 1971, Idris, still an Air Vice Marshal, was awarded the Param Visisht Seva Medal. It was, I learnt, the very first time that it was awarded to an Air Vice Marshal, as earlier it had been only for Air Marshals or for the Chief himself.

One day, months later, the CAS, Air Chief Marshal Lal, called Idris to his office and asked him if he would take on the assignment of being the head of Indian Airlines. As it was urgent, the Chief wanted Idris to respond by the following morning. I was

Idris receiving the Award from President V.V.Giri

Idris and his mother with Prime Minister Indira Gandhi at the Award ceremony

in Bombay on a visit to my mother who was not well. He phoned me to ask me what my reaction was to this offer. I immediately said that it was completely up to him to choose where he would be happiest. It is the man who does the work and shoulders the responsibility, so he must be free to decide what he would find more fulfilling. I would be happy, I said, with whichever work he found more congenial and satisfying. Idris, without hesitation, told the CAS the next morning that, provided the Chief had not already made any commitment to the government, he would much prefer to stay with the Air Force. Idris never regretted his decision. With his next posting as Air Officer, Administration, he was given the rank of Air Marshal.

During the children's summer holidays, Idris managed to get a little holiday time. Amongst these memorable holidays, one year we went to Kulu Manali by road, with Idris driving all the way. We stayed in Dak Bungalows and occasionally at a Military Engineering Service Inspection House, and often did our own cooking. We visited lovely places, snow-clad regions, apple orchards and drove along rivers and the old wooden houses up in the hills which were very charming. We were even able to witness the local dances with the dancers in their colourful ethnic costumes. Another wonderful trip was to Kashmir, which really is the Switzerland of the East. Even in Delhi, if we could get away for a weekend, we went to little places near remote villages and stayed at Dak Bungalows or Inspection Bungalows near forests or rivers. It was always wonderful for the children and for us to be able to get away like this.

In spite of his long hours, Idris found his assignment as ACAS (Plans) a most challenging one, as it entailed the responsibility for the complete growth and development of all key elements of the Air Force. I remember how his mind was always on this task even after the long hours at office. His farewell speech at the end of this assignment was as follows:

Farewell address by AVMI.H. Latif, PVSM on relinquishment of appointment of ACAS (Plans)

First of all, may I say that I welcome very much this opportunity of saying a few words. The Air Force has been very good to me and has given me very fine appointments. In fact, I can truthfully say that there are very few, in fact none, that I have not enjoyed. Perhaps it is because I, and indeed my wife and children, also love an Air Force environment; but of all the appointments

that I have held, none have I found more stimulating and challenging than that of ACAS (Plans) — stimulating not only because of the type of responsibilities that went with it, but, equally, if not more so, because I have had the pleasure and privilege of having as my colleagues in the branch hand-picked and dynamic officers, officers full of ideas, full of initiative and enterprise; and also, if I may say so, with tremendous capacity for sustained hard and dedicated work. Very often, I had no option but to call upon many of you to accomplish a task within a seemingly impossible time-frame and it was proved to me over and over again that, in our Directorates at least, the guiding principle has been the well-known phrase:

THE DIFFICULT WE DO AT ONCE; THE IMPOSSIBLE TAKES A LITTLE LONGER!

It was a new branch to which all of us walked into, a new branch with no guidelines except those of our own making. Whether our ideas were right or wrong, time alone will tell. But I do believe that all of us together have evolved a system that achieves something tangible in the wide and varied field of planning.

Some years ago, when I came across the following illuminating article, I hoarded it — like so many of my own notes and paper clippings, I felt that I just had to add it to this book. It is a copy of an article by Commodore Ranjit B. Rai (Retd.), former DNO and DNI at NHQ, an IAF-trained Aircraft Controller and author of A Nation and its Navy at War (Lancers, 1987). The article was in the Deccan Chronicle dated 31-3-2007, and was about our Air Force's role in setting aflame the oil tankers in Karachi during the Indo-Pak War in 1971.

IAF Strikes Karachi Oil Facilities

India's Armed Forces saw their finest hour in the 1971 war against Pakistan. The IAF's contribution got side-stepped. The late Air Chief Marshal P.C. Lal alluded to this strike in his memoirs, but only in passing. Yet the facts deserve recalling before amnesia sets in. This writer's revelation is not meant to ignite an inter-Service fire, but history, after 35 years, deserves to be recorded as it was, without rancour.

At dusk on December 3, the Pakistan Air Force had struck all IAF airfields on India's western border and caused widespread damage. The IAF could not respond as its fighters did not have night capabilities. Just after midnight, Mrs. Indira Gandhi broadcast, "The war on Bangladesh has become war on India. We have no option but to put our country on war footing". In the wee hours of December 4 morning at Jamnagar, OCU head, Wing Commander Don Conquest and his highly qualified pilots who had recently inducted Hunters Type 56A and 235-gallon drop tanks, learnt they had no assigned role for that morning. Conquest approached his OC, the legendary Air Commodore Pete Wilson, and insisted that his boys were capable of a strike on Karachi. "What will our wives and children who have been shunted out from the base's Bhangi Barracks to Jamnagar town say?" was his plea. "That we stayed on ground"?

Pete Wilson, who was busy planning and readying the MiG-21s to hit Badin and Drigh Road with first light, let Conquest plan his mission. Eerily, on December 4 morning, four OCU "Top Guns" took off for "target Karachi" with road maps. The Hunters could not carry rockets, as two 235-gallon drop tanks were slung on the pods, to ensure the reach at low level and just five minutes over target. Their sole weapon was their 20mm cannons. Wing Cdr. Don Conquest, the strike leader, is now settled in Australia, Squadron Leader S.N. Medhekar his Winger lives in Pune, Flight Lieutenant P.K. Mukherjee resides at Salt Lake City in Kolkata and Flight Lieutenant S.K. Gupta in Delhi. All have recalled their mission many times over.

Don Conquest recalls. "As per SOP (standard operating procedure) off Karachi we dipped our noses and fired a few rounds into the sea to test our guns. Mukherjee's guns had jammed, so the three pressed on at 500 ft. along the coast. As we neared Karachi, the large oil tanks loomed out of the sky, their silver paint shining in the sunlight. We made two runs without difficulty and after the first, there were huge balls of fire and volumes of smoke coming out of the storage. As the smoke haze made flying dangerous, we aborted the other runs and flew back".

On landing back at Jamnagar and before he could file reports and process the films, Conquest was ordered to fly his OCU to Jaisalmer. Indian Army

tanks were under siege. The oft asked question is why the IAF has never made much of this amazing achievement.

When this was brought to the notice of late Air Chief P.C. Lal by Flt. Lt. P.K. Mukherjee (later Air Vice Marshal), Lal magnanimously said, "Let the Navy take the credit". This revelation in no way detracts from the brilliant and bold planning by the then CNS, Admiral S.M. Nanda and his staff, and the superb execution by his commanders and captains at sea.

So ended the article, but interestingly, many years later when Idris was in France from 1985 to 1988 as Ambassador of India, he happened to meet an old school friend of his who was on a holiday in France when we bumped into him at a formal reception. He, in a small way, perhaps, confirmed what Rai has recorded. Idris later told me that after they exchanged warm and affectionate greetings, they were chatting away catching up on the news over the years. This friend, Nasiruddin Khan, had earlier been the General Manager of Burma Shell in Karachi during the period of the 1971 War. After all the greetings, he said, "Idris you owe me a bush shirt," and went on to say that he had been on his way to the oil tanker facility on 4th December 1971 for an official inspection when the four Indian Air Force Hunters roared over. He said that he dashed out of the car and hit the dust, and just lay there as the planes flew on bombing the oil tankers. Soon the tankers were ablaze.

As Commodore Ranjit Rai has written, it was an amazing feat of bravery to have attacked — particularly in broad daylight — such a key target which, of course, was, as a consequence, heavily defended. The raid, incidentally, was carried out soon after dawn by the pilots of Operational Conversion Unit (OCU), operating from Jamnagar. It was recognised as one of the 1971 War's most daring missions.

Air Force Wives Welfare Association: Helping Hands

After Idris got posted to Delhi as Assistant Chief of the Air Staff (Plans), we first lived at Lodi Road and then were allotted 6, Safdar Jung Lane, just a few houses from the Prime Minister's residence. Our cup of joy was full and overflowing, for our three children could stay with us now at our new home in Delhi.

Having been informed by Mrs. Lal, the Founder President of the Air Force Wives Welfare Association (AFWWA), that I had been elected in absentia as the Honorary

Secretary of AFFWA, I had to get down to working out the rules and regulations, aims and policies which needed to be considered and formulated for AFFWA and its membership. It required funding as well as systems for effective interaction between the central body, the Commands and stations. All this had to be worked out in consultation with the general body. An office and officer for each AFFWA branch in every station was also required, for what could AFFWA do without this crucial liaising, and help from the Air Force Stations and Headquarters? Urgent action included the rehabilitation of widows, of providing them with training in income-generating skills, if needed, and housing, if possible, even for a while. Their children needed to continue their education in good schools. An ethos of caring and of working together towards worthwhile goals had to permeate all the Air Force families in our stations all over the country.

Soon after, AFFWA's involvement went beyond the Air Force when the nation faced a great challenge as millions of refugees from Bangladesh poured into India where they were given whatever aid and shelter was possible. Large amounts of basic requirements for them included clothing. AFFWA got down to collecting these from all over the country. The clothes were kept initially at the old Central Vista Officers' Mess, and different groups of AFFWA ladies would come in turn to help me go through the bundles to see that the clothes were in reasonable condition. I would load up our car with bundles of clothes and drive off to deliver them to Mrs. Lal at Air House.

One day, Mrs. Lal looked at me with concern. I was wearing a white cotton sari, with no jewellery or bracelets. Then she spoke to me with gentle kindness just as if she were my older sister, and said, "Bilkees, just look at yourself. You must at least wear bangles, I feel scared for you when I see you like this"! What she meant was, a resemblance to a traditional widow, and she cared enough to point it out. Long afterwards, Mrs. Lal once heard that I was not well. She was not in Delhi but she telephoned me long distance to tell me not to neglect myself and gave sound advice to see that I would do whatever was required to get well again. This kind of caring inspires one to want to do more for the common good.

I was asked to continue to be the Secretary of AFFWA after Mrs. Satya Mehra succeeded Mrs. Lal as AFFWA President. Mrs. Mehra was also very warm and affectionate, and also a very caring and deeply involved President. She proceeded to establish the AFFWA shops and stepped up vocational training at each Air Force

station. She was also very concerned about those wives who lived at Air Force Stations in remote areas. No jobs were available in such places even for those who were well-qualified. She felt that the newly set up AFFWA shops at the stations could include and sell attractive or useful items made by AFFWA members. This would give them an opportunity for their creative talents and skills to be utilised, and become an outlet that would enable them to have a source of some personal income. The AFFWA shops became very popular and have continued to also provide AFFWA with funds for welfare purposes, with a variety of goods available at great savings to each buyer.

Mrs. Satya Mehra and Bilkees

Mrs. Mehra said that we should plan to hold an AFFWA Mela in Delhi. For this, items from different Commands would be brought to Delhi as each would set up their own stalls. Goods from the AFFWA shops of each station could, thus, be displayed and sold. However, for me as the Secretary, it involved detailed organising and coordinating with our Command AFFWAs. A few weeks earlier, there had been a huge Railway wives Mela in Delhi, and several big schools had also held fairs. At

the next AFFWA Board Meeting, I showed newspaper cuttings, pointing this out to Mrs. Mehra, and suggested that we could drop the idea of holding a Mela in Delhi as several similar ones had recently taken place! Mrs. Mehra, is a most understanding and kind lady and with deep spiritual faith which is very inspiring. She laughed and told me that it would be good for the AFFWA wives to have this opportunity to sell their items, and it was also good to publicise AFFWA and its work. She was right and her smile and gentle laugter made me then enthusiastically go ahead and put in my full effort for the Mela.

Mrs. Satya Mehra never held it against me that I had suggested that we should drop her plan to have a Mela, and always remained warm and affectionate. This is very important for organisations whose members have to work together, because people must be allowed to disagree without facing disapproval or else those in charge may never get to know the true opinion of others. However, on the other hand, dissent must be made with politeness always, and be well meant, and should not be voiced with rancour or anger.

I then got going on organising the Mela and coordinating with all the numerous Command AFFWA groups. I went to see the authorities concerned for the use of the grounds of the very centrally located and beautiful Hyderabad House for three days. They agreed on condition that AFFWA would leave the place neat and clean when it was over, especially as they had an important function for a foreign dignitary the following day. The day arrived and the Mela was very well attended and was bustling with people and activities. The whole experience taught us all how to improve. It also brought about a great awareness and feeling of togetherness between AFFWA members from Commands all over the country. I still remember the night when the huge Mela was over, and I was sitting there all alone on a grassy mound at 2 o'clock at night, overseeing and waiting to make sure that after the stalls were dismantled, each piece of paper, food stuff, paper plates, and all kinds of garbage and debris were all cleared up before I drove myself home! In those days, one could drive home alone quite safely even at that hour!

CHAPTER 18 Air Officer Administration and on to Central Air Command

Early in 1974, Idris was promoted to the rank of Air Marshal and appointed Air Officer for Administration at Air Headquarters. We packed and moved to our newly assigned accommodation, the stately house and garden of 6, King George's Avenue, right next to that of the Chief of the Army Staff.

A few weeks later, the children asked Idris if he was finding his work rather boring, particularly after the challenging tasks that had earlier been entrusted to him. Idris' immediate response was that he certainly had thought that it wouldn't be exciting — or even very challenging. But, he added, the Air Officer, Administration, was also responsible for all personnel matters, and in the armed forces, the morale and motivation of the men — be they officers or the non-commissioned ranks — have always proved to be amongst the most critical factors. It was clearly for him not only yet another challenge, but emotionally also a most rewarding one. In his initial briefing to his officers, he told us, that he, therefore, took particular care to stress this. As an example, he said, that he had mentioned that this was important when one receives an application for help from persons facing acute personal problems. Dealing with these, even under pressure of time, one must never, ever just glance through and dispose of the application. For those in authority, it may be one of many, but for the applicant, one must recognise that it may well be very serious and he may be desperately hoping for our understanding and our help. It must also be kept in mind that many may not be able to express themselves too well in English.

Responses, in turn, must, therefore, never be just a brief two-line letter, but should be drafted carefully and with great sensitivity. By far the best way, he had specially

Air Chief Marshal O.P. Mehra

emphasised, was to somehow find time to interview the person, talk with him across the table, particularly if the seriousness, one felt, justified it. Often, very often, he said, he himself had found that even for one's own understanding of human problems, a personal hearing gave a truer picture of the problem; and, a satisfactory solution became that much easier to identify. At the very least, the applicant better understood Air Headquarters' constraints, and almost always left with a feeling that to resolve his problem, a very personal interest had been taken at a high level. To set an example, he concluded with a laugh, whenever needed, he practised what he preached! As a result, not infrequently, he started looking at other files only at 5 in the afternoon.

To Central Air Command

Within just four months of assuming charge, Air Chief Marshal Mehra, the Air Chief, called Idris and told him, "Idris, I know that you have been Air Officer Administration for barely four months, but I really feel that, for your own good, at this stage of your career, you should have an Operational Command under your belt. As it happens, Central Air Command has fallen vacant and I want you to take over as soon as possible. I know that yet another move at such short notice will be very trying for the family, particularly for Bilkees, to move house yet again and, of course the children. But I am taking a long-term view of your own future. Remember, an Operational Command is an Operational Command, and very, very few Air Marshals get such an opportunity." Idris was naturally thrilled but also very conscious that, domestically, it would be tough on all of us, especially for the two older ones, Mariam and Asad, who were now quite grown up and about to face a rather sensitive period of their lives. Clearly, therefore, it was inadvisable for the children to come with us to distant Bamrauli (near Allahabad), where the Headquarters of the Central Air Command was located.

Asgar had to stay back too as he was then in his last year of school, and had to become a boarder at the Air Force School at Subroto Park in Delhi, where he was studying.

To their great credit, the two older children told us not to worry about them, and entirely *on their own*, and within days, were able to find acceptable and interesting jobs for themselves. Mariam, after graduating in English Honours from the Lady Sriram College, had thereafter just completed her two-year Teachers Training degree, having come 4th in the whole of Delhi. She was very keen on travelling, and applied for, and got selected as, an air hostess. Asad, who had graduated from St. Stephens in Economic Honours, joined a good firm within the week and found himself accommodation with bed and breakfast at the home of a close friend. The friend, by a strange coincidence, was leaving India, and his mother kindly accepted the rental room and breakfast arrangement with Asad. Having to leave the children behind was, of course, emotionally, not easy for both of us. I was deeply saddened, but we realised that there was no option. Having just finished setting up house barely a few months earlier, with the garden at 6, King George's Avenue looking unusually good, but like so many others in the Defence Services, I got down to packing again and, within a few days, we moved to Bamrauli. My mother had shown me how to pack glassware carefully when I got married. She herself had brought a full set of beautiful cut glass glasses with her from France when she got married, none of which had broken or cracked. After all our own moves, safe packing was not a problem.

After our arrival in Bamrauli, in the early days, Idris would join me every evening whenever possible, and we would slip away to the terrace upstairs to watch the sunset and, in our thoughts and our prayers, for a while at least, be with our little family in far off Delhi. Once though, I remember that I was rightly reminded by a young friend that I should be happy for Idris who, with this posting, had been entrusted with a key operational Command, and luckily, at a critically important stage of his career. This was true and I was grateful, but, nevertheless, one but naturally missed the children. However, there was a great deal to be done at the Command Headquarters, and we found our Air Force families to be a wonderful group of people. Allahabad was also such a historic city of very friendly and cultured people, and we very soon were greatly drawn to them. Having moved by then, to 17 different living quarters, like most armed forces wives, I had also acquired some expertise in either lengthening or shortening our curtains according

The House for Air Officer Commanding Central Air Command,
just off the runway at Bamrauli Air Force Station

Idris taking over Command of Central Air Command from Air Marshal Krishna Rao

to the heights of the doors and windows in the latest accommodation. Our house at Bamrauli was unusually large, airy and the rooms were enormous. The drawing room soon looked lovely and we entertained a large numbers of guests to meals throughout our stay there. There was a garden in front in which we were thrilled to be able to grow lovely flowers, and even strawberries as the climate and soil proved conducive! The house was right next to a hangar and not far from the main runway, but we soon became accustomed to the sound of the Packet aircraft revving up and testing the engines every morning. The Packet engines were tested at 5 am to make them ready for a crack of dawn take-off. Competing with this were the pre-dawn sounds of prayers from holy places in the vicinity, especially from a Gurdwara nearby, which proved to be very soothing.

As the Commander-in-Chief of Central Air Command, Idris was now directly responsible for two major and key operational elements of our Air Force: its Bomber Force equipped with Canberras at Gorakhpur, Poona and Agra; and also a major element of its Transport Force equipped with the Fairchild C-119 Packet aircraft, part of which was based at the Central Air Command Headquarters at Bamrauli. This, to Idris' delight, provided him an excellent opportunity to do his conversion on it and also, thereafter, gain first-hand experience in the supply operations over the Himalayas. Often, Idris told me, these were at rather remote and challenging landing grounds, located at unusually high altitudes, such as Thoise in the Ladakh region. But, for Idris, I soon realised, it provided exciting opportunities to hone his skills on, what for him was yet another type of aircraft! These and such transport sorties, had been for decades, I learnt, the lifeline for all our Army and civil engineer formations deployed along our northeastern front. This, for Central Air Command at that time, stretched from Ladakh in the north, through Sikkim right on to NEFA (Northeast Frontier Agency) in the east; and right down to the Car Nicobar and Andaman Islands. Idris, who had spent most of his career flying fighter aircraft, was delighted to have the opportunity of continuing to fly our largest four-engined Super Constellations based at Pune (still, during that period under Central Air Command), aircraft on which he had already converted when he was commanding No. 2 Wing.

Fairchild C119 Packet

Thoise in the Himalayas

On arrival at Bamrauli, where the Headquarters of Central Air Command was (and is) located, one of Idris' high priorities, as I have earlier mentioned, was to do a full conversion on the Fairchild Packet, the backbone of the Command's transport fleet.

Within a couple of months, he decided to fly to Thoise, an advanced landing ground in the northern Himalayas. The single runway there was not only short, but also not at all properly surfaced, and, therefore, considered a particularly difficult one on which to safely land. Half jokingly, his co-pilot told him that if, after touch-down and braking, the tyres smoked (almost inescapable because of its height and very limited length) indicating that the pilot had slightly overshot and had to brake harshly on this short runway, it was customary for the captain of the aircraft to stand drinks all round! Idris says that by sheer luck, and certainly to his surprise, he made a good landing — at least, that is what his polite co-pilot told him! "There was some smoke, Sir, but, a *very* limited amount, and of an acceptable level". To his crew and others at the landing ground, he said, that though the smoke, he had been told, was hardly there, he wanted to celebrate his first landing at Thoise and announced "a double beer all around"! It was well-known in the Air Force, of course, that Idris never took to alcohol (because to him it has, whatever the quality, a foul taste!). For this occasion, however, he felt that he should take at least a few token sips, and pretend that he enjoyed it — then staying on for an excellent lunch!

It was after this personal assessment of the inadequacy of the runway and the support facilities available, that, within days, he specially visited Air Headquarters, to personally pursue the case for the runway to be not only lengthened, but properly surfaced; and the very small care and maintenance party upgraded into a full-fledged advanced operational base. He deliberately first met the Director of Operations (Transport), who had earlier worked under Idris and proved his competence. He wanted to meet him first, he said to me, to personally find out from him, and fully understand the reasons for his reservations. Thereafter, Idris took it up with the Vice Chief of the Air Staff and, finally, the Air Chief himself. To his great delight, both wholeheartedly accepted Idris' proposals. The Air Chief, however, rightly added, "Now we have to somehow convince the Ministry to release the funds!" Thoise is now said to be a well-developed forward base of our Air Force, and crews are able to fly in and out with the requisite degree of safety.

At Thoise in the Himalayas

Canberra Force

Many years later, after he had retired, and the Canberras were being phased out, he narrated to me something that I know must have troubled him greatly as it, he felt, seriously affected his relationship with a person whom he and I knew well. Idris said that soon after taking over command of Central Air Command, in terms of operations, he identified a challenge which awaited him and which he felt was of prime importance: this concerned the utilisation of the Canberra bombers in the event of another conflict. Fortunately, he found that not only his

senior operations staff, but also every Canberra Squadron Commander — most of whom Idris knew well — fully shared his view. In the earlier conflicts of 1965 and 1971, as is well known, the Canberras being obsolescent (if not obsolete!) were, in the interest of survival, almost invariably tasked for 'nuisance' night raids by solitary aircraft on distant military targets of special significance. Such raids, Idris, with the others, strongly felt, even at this late stage of the Canberra's life, did not do justice to its potential, particularly in our environment. The officers were all, without exception, strongly of the view that in previous conflicts with our neighbours, the Canberra, the only bomber on the inventory of our Air Force, though rapidly getting outdated, was still all that the Air Force had that could, even individually, with its sizeable bomb load, inflict substantial damage.

At the next Commanders Conference, therefore, Idris proposed to the Air Chief, Air Chief Marshal Mehra, that, since our Air Force had no other attack aircraft with a substantial bomb load, a much more meaningful task for the Canberra force would be to utilise them in low-level daylight raids, preferably at dawn or at dusk, against the several key targets, that were within 40 miles or so of our border. He remembers stressing that at the very low level possible in daylight, coupled with the very short distances over hostile territory, surprise could be ensured, greatly minimising the risk of interception. This would enable the Canberras to accomplish their mission with high accuracy, and still get back with an acceptable chance of survival. At the conference, there was opposition but, happily, after much debate, the Chief accepted Idris' proposal with the proviso that the validity of what Central Air Command (CAC) had recommended, be first tested by an exercise in which Western Air Command would be responsible for the defence of the targets, while the Canberras of Central Air Command would be the attacking force.

A month later, such an exercise was initiated by Air Headquarters in which 20 targets were to be attacked in daylight by the Canberras. In the exercise, attacks on no less than 17 were successfully accomplished; while only 3 could be intercepted by Western Command. Based on this, at the next Commanders Conference, the Chief ruled in favour of Central Air Command. Idris and all of them at Command Headquarters, as well as the Canberra crews were, understandably, jubilant.

Flood Relief Operations

Heartwarming and unforgettable events whilst Idris was at CAC, were the Patna and Orissa flood relief operations undertaken by the Command. He especially recalls with both pride and happiness, the role of the ladies (the wives of our Service personnel), who spontaneously came forward to help me by cooking and packaging thousands of packets of food for air-dropping by our splendid and courageous helicopter crews — both pilots and the ground crews. Some pilots, on more than one occasion, were flying no less than 20 sorties in a day. I also remember that, to thank us all, our then Defence Minister, Shri Jagjivan Ram himself, flew down to congratulate the Air Force detachment on its splendid performance, and, very specially, the ladies and me, for the exemplary spirit of service we had shown.

The large compound of the Patna College served as an advance helipad which greatly helped to minimise flying time and maximise relief sorties. This enabled some of our pilots, as mentioned above, to occasionally accomplish no less that 18-20 sorties a day. Incidentally, we had also taken our personal cook along to set up a sort of canteen for quick cups of tea and snacks between flights for the helicopter crews. On one flight on which Idris specially asked me to join him, we saw two nuns standing on the roof of a half submerged building holding up a sign asking for an emergency supply of powdered milk and bottles of drinking water for over a dozen children who had been trapped there. Idris and our crews ensured that, within the hour, these were air-dropped.Three days later, Idris received a very movingly worded postcard from the nuns thanking them all for the exceptionally prompt response.

Apart from Patna in Bihar, there were devastating floods that year in UP and in Orissa also; and the interaction of his officers and crew with everyone, the government officials from the Chief Minister downwards, and others who had volunteered to lend a helping hand, was no less satisfying and heartwarming. It was particularly touching when, in spite of the very late hour of Idris' arrival, the Chief Minister, Shrimati Nandini Satpati herself, met him to underline the importance and urgency of relief. She also ensured that the state administration's participation was total and that everything needed was provided on top priority. To top it all, when this was accomplished, she hosted a very special private dinner to thank Idris and the Air Force; a dinner to which the heads of the affected villages had also been invited.

A visit by the Defence Minister Shri Jagjivan Ram during flood relief operations, Patna

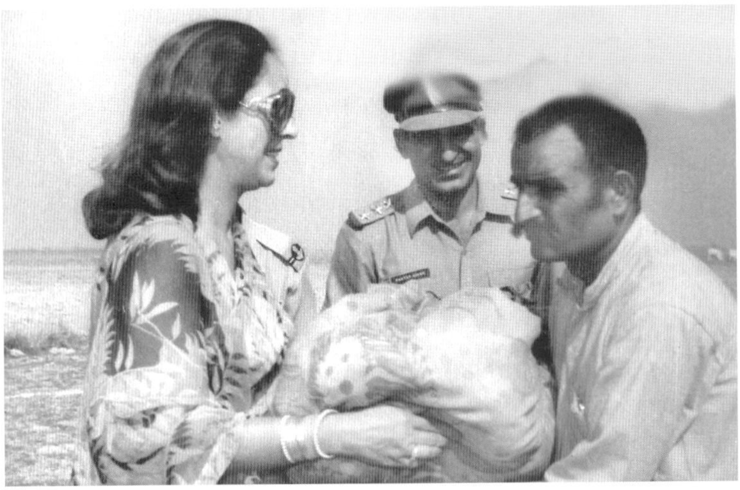

Bilkees during flood relief, Bihar

Colour Presentation Ceremony

Towards the end of our posting, the Central Air Command was given the responsibility of organising and hosting the Presidential Colour Award Ceremony at Agra, then under the able and sensitive command of Air Commodore 'Timky' Brar. Idris had a discussion with him on the general plan that he had in mind and, during the preparatory weeks, made two short visits to sort out any problems that the AOC (Air Officer Commanding) may encounter. On the final day, on Idris' arrival in time to receive the guests in Agra, we found the station looking absolutely immaculate. As planned, to mark the commencement of the ceremony, a perfectly timed, low-flying Canberra made an exceptionally accurate air-drop of masses of rose petals, barely a hundred metres beyond the front line of the spectators. The thunderous applause of the spectators that followed naturally thrilled us all. Seeing Air Commodore Brar's happiness and joy (and relief, I'm sure) rather moved Idris and me, because Idris knew that Timky Brar and his team of officers and men had worked long and hard. The Air Chief, before he left, was gracious enough to tell them all that it was, in his view, the best Colour Presentation that he had ever witnessed. Within an hour of our return to Bamrauli, Idris took special care to send a congratulatory signal to the AOC, Air Commodore Brar. The latter told us later, that when he received it, he read it out to all those at the party that had been organised in the Officers Mess that evening to celebrate the successful conclusion of that major event.

About a year later, as numerous personnel of Central Air Command received a number of annual Air Force awards, I have included some of Idris' thoughts which he expressed in his address when the awards were presented.

> *It has been my proud privilege to recognise this morning some outstanding personnel of this wing, some commended by me and some by the CAS himself. I congratulate them and wish them continued success. However, I would like all of you to know that I am also very conscious of the fact that whatever the units/depots have achieved, whatever the Air Force has achieved, the laurels that have been won in the field of battle, or the kudos that have been earned, by the manner in which we have accomplished our normal peace-time duties and responsibilities — all these achievements are*

not only by the efforts of the few who are singled out each year for awards, but the dedicated hard work of many, many more who are nearly if not equally deserving of recognition.

Almost a full year has passed since I last addressed you — a year in which there have been substantial achievements by this wing, the Command and, indeed, the Air Force as a whole, achievements which have reinforced and, indeed, enhanced the reputation that we, as a Service, enjoy.

The Flood Relief Operations are one striking example: a major responsibility fell on this wing and practically every element was called upon to play its part. The manner in which you discharged your duties won praise at all levels and also, of course, the undying gratitude of those whose lives and property were saved from being devastated.

Flight safety is yet another area in which an appreciable degree of success has been achieved by our Air Force. It hasn't made banner headlines, but obviously it is of the most vital consequence. During these months, the earlier alarming increases in accidents have not only been halted but the trend reversed. The instructions and directions that those at Air Headquarters issue are undoubtedly most important, but without doubt, the credit goes primarily to the stations and the squadrons, the air crew who fly our aircraft and the ground crew who service them, and also various essential support services. In the coming months, let us continue to do our utmost in this direction. We certainly can afford no complacency.

Last year, I spoke to you about the care and maintenance of our equipment, in particular, our aircraft and the other sophisticated weapon systems that the country has procured at very great expense, expense that the country can ill afford. Is your conscience satisfied that everything that can be done in this field is, in fact, being done?

There is one final point that I would like to stress. The strength and capability of our Air Force — of any force for that matter, depends not only on what we have, but much more on what we do with what we have. The long history of war has proved, if nothing else, the supreme importance of

the human element in the total weapon system. What ultimately decides the outcome of any battle, of any war, is the quality of the man — his training and his morale. Happily our Air Force in both strength and sophistication is far better off in comparison to our potential enemy. If our thinking and our training remain positive and progressive and operationally-oriented, we can be invincible. Let us resolve to leave no stone unturned to make it so. In this, as in all else, I can think of no better guide than that given by a great statesman: let us not ask what the country can do for you, but what you can do for the country.

The audience was very obviously deeply appreciative and moved by what Idris said to them.

I learnt from a particular incident at Air Force Station, Bamrauli, the Headquarters of Central Air Command, about the importance of activities wherein ladies could participate and, thus, also bring people closer together. At this Command Headquarters, I found myself very busy as there was always a great deal of station involvement with medical, schooling and other welfare needs. We held competitions for Better Homes and also Best Garden for the Senior Non-Commissioned Officers (SNCOs) and other ranks which were popular, and the AFFWA shop was doing well. However, one day I was told that the wife of one of the officers was quite unwell and in hospital, so I went to see her there. I found this young lady to be in a deep state of depression and on the verge of a nervous breakdown. While I was with her, the lady broke down and sobbed and then told me that she was in a terrible state and that she had been feeling very lonely and isolated. She said that she and her husband had not been at the station for very long, and she didn't know too many of the others at the Air Force camp where we all lived. She added that her husband who was a good and kind man, was extremely busy at work, and when he came home he was so tired that he would just sit and read the newspaper. She was more or less on her own, with plenty of time on her hands. She told me that she was also upset while at the hospital, in case people might think that she had broken down because her husband was harsh and cruel, whereas actually he was a very caring man! She was really in a bad state and I realised, therefore, that we just had to involve such young wives of the officers in interesting station activities.

Soon we organised AFFWA cultural shows and also plays in which even some of the husbands took part, in addition to the other activities and games. All the shy ones were brought into these activities somehow and soon there was a happy atmosphere, among young and old who became like a great big family group! The depressed girl turned out to be an excellent amateur actress and also wrote good poetry which got recognised. Soon she was a star in the plays and was all smiles and full of enthusiasm. Those who joined in the activities found that they had far more fun, more of a sense of belonging to a vast family with sharing and caring.

In my own case, I often recalled when I was a very young, recently married wife, at the Joint Services Wing at Dehra Dun where Idris was posted. I was completely shaken when the wife of the Commandant of the Wing suddenly informed me at a Ladies Meeting that I was to be in charge of the arts and crafts stall at a forthcoming fair. Because I myself had found it so tough at that time, therefore, I had never tried to push anyone into activities thereafter. Nevertheless, one should have a sense of caring for those in need, and participate in bringing about improvements wherever possible. I did feel, however, that there is a sense of belonging in this big Air Force family for usually we all live together, celebrate festivals and joys with Air Force neighbours and friends. In times of tragic deaths and a shattering loss of a dear one, one is surrounded with the loving care of these Air Force authorities and of AFFWA members, and friends who help in every way. One's real families are far away from Air Force Stations, and this closeness often lasts and continues through the years even after retirement.

We travelled to other Air Force Stations under this Command and Idris was happily able to do a lot of flying. Our visits to Ladakh and the Andaman Islands for one of our units at Carnicobar, under Central Air Command during those years, were particularly memorable The Andaman Islands have especially beautiful beaches with the sea a clear blue, near the Air Force Station at Carnicobar, one of the larger islands of this group. What fascinated us were also the people and the way that they handled the law and order of their villages. The police official of the government who was posted there, actually grumbled at his lack of work! The villages were at a little distance from each other, each with its own elected "Captain" in charge, with a Panchayat type group with him to resolve law and order problems. We were told about their revered but late leader Bishop Richardson, who had been the supreme leader as well as the religious leader as the people had become Christians. The people of the island had looked up

to him for years, and he had brought about peace and unity amongst the people. They had suffered greatly earlier, during World War II, because of the brief but greatly resented Japanese occupation. In Carnicobar, we were told that their occupation also resulted in numerous island women being infected with sexually transmitted diseases which were very traumatic and spread rapidly to others, as no proper medical facilities were available at the time.

The Bishop was no more, but we made a formal visit to the Vice Chief, who was a most remarkable personage. We had been told that when the Air Force was establishing its station in Carnicobar, Pandit Nehru had told the Air Chief that the tribals must be treated with utmost fairness and no advantage taken of them in any way. Panditji had emphasised that any land acquired for an Air Force base must be adequately paid for. Accordingly, the Air Force officer in charge at the time, went to visit the Vice Chief of Carnicobar as he owned vast tracts of land. The Air Force had found these suitable for a runway for our aircraft to land and take off from, and also for the other requirements of land for the Air Force base to be set up in Carnicobar. They requested the Vice Chief if they could buy the required land from him. He agreed to give it, but he refused to take any money even though it was for a vast amount of land. When they insisted and told him what Panditji had said, he replied that the only payment he would like to have was either Pandit Nehru's cap or his jacket! As he was very firm about this, Panditji sent a jacket which was brought for him, and the required land for the Air Force was handed over! It was a huge area that he so generously gave and was enough for the whole Air Force Station. When Idris and I went to pay our respects to him, we reiterated how incredibly gracious he had been. He was an impressive, tall and well built, jovial gentleman and he welcomed us with great warmth. He pointed out to us Panditji's jacket in his living room which was mounted and kept in a place of honour, and he said that it was his most precious and prized possession.

The Air Force base at Leh, the capital of the Ladakh region, was another memorable place in Central Command that I was privileged to visit when Idris took me when he went there. This very unusual place, with its stark and bare mountains and little or no greenery, is located in a valley, and flying out seemed to me to be a really challenging and even risky task because of the very high mountains, and narrow valleys coupled with the extremely strong winds. But, the people of Ladakh looked most attractive, especially with their beautifully coloured costumes and high fur hats. The monasteries

on the hills also looked so serene and beautiful that one would always want to return. I felt like staying on there much longer, but Idris had to fly back once his work was done.

An interesting and different kind of function that we were involved with was an initiative by 2 Wing in Pune which organised a grand show with the help and support of Nari Hira and film stars of Mumbai. The chief guest was the Governor, Nawab Ali Yavar Jung, my father. Idris' rather entertaining speech on the occasion is given below.

ADDRESS IN MUMBAI WITH NAWAB ALI YAVAR JUNG, THE GOVERNOR OF MAHARASHTRA, AS THE CHIEF GUEST, AT A FUND RAISING ENTERTAINMENT FUNCTION

Your Excellency, Nawab Ali Yavar Jung, Air Chief Marshal Mehra, Mrs. Mehra, distinguished guests, distinguished artists, ladies and gentleman. Standing before you here this evening to say a few words of introduction, I am reminded of the occasion many years ago when I was in school and we were being lectured on the subject of elocution. "If you have to speak", advised our professor, "try always to fashion your speech like a woman's skirt — long enough to cover everything but short enough to be interesting!" Benjamin Disraeli, that great British statesman and parliamentarian, said much the same thing except that he went further. A young enthusiastic parliamentarian, who had just been elected and was extremely keen to do well, sought Disraeli's advice. "Sir", he said, "I want to start off on the right foot. How often do you think I should get up and speak?" Disraeli is said to have replied, "Young man, I suggest that you do not speak at all! It is better that the audience should wonder why you never spoke, rather than wonder why you ever did!

With such thoughts in my mind, I promise that I shall be as brief as decency will permit.

What is perhaps one of the greatest tragedies in the life of any nation for the soldier, the sailor and the airman in the aftermath of war is, in some ways, more tragic, for it leaves behind a trail of broken homes and broken families. In the Air Force, the occupational hazard is perhaps the greatest,

as many casualties occur even in times of peace. The responsibility of funds for the shattered widow and helping in the education of the children is largely ours. Apart from the ravages of war, Service personnel for several reasons, have to retire at a relatively young age, often in the 40s, always by the mid-fifties. They have to retire at a stage when the financial burden of supporting their families, especially educating their children, is the greatest.

This great gathering here, the presence of the large and distinguished audience, a fine cross-section of the people of this great city of Bombay, to us in the Air Force is symbolic of your support, a symbol of your never failing sympathy and your constant concern for the welfare of those in uniform. Most of all, Your Excellency I would like to thank you and Begum Ali Yavar Jung, on behalf of all ranks of Central Air Command for honouring us with your presence here. We are very conscious of the demands that are made on your time and the pressure under which you have to function, and we are really most deeply grateful to you for having found time to be with us on this occasion. I would also like to thank Air Chief Mehra and Mrs. Mehra for flying down from Delhi to be with us. Their deep concern and the personal interest that they have always taken in the welfare of the Air Force family is common knowledge for all who have the privilege to serve under his flag. Always, we have been able to lean on them and count on their unqualified support. Indeed, it was with an appeal under the Air Chief's signature that we launched this project.

I would also like to thank the glittering galaxy of stars, these dazzling friends of the world of entertainment who have so spontaneously come forward to enrich our programme of entertainment for you.

We owe a very great debt of gratitude to Mr. Hira, that young, handsome and dynamic gentleman who has indeed been the chief architect of the lavish entertainment which you will witness this evening.

And, finally, I cannot help but thank Air Officer Commanding Maritime Air Operations (AOC MAO) and his very able and dedicated band of officers

and men, all of whom have worked so hard, for weeks and months, to make this function possible.

Too soon, it was time for a move from Central Air Command. The new Air Chief posted Idris out to a non-operational Command as Air Officer Commanding Maintenance Command, and Idris was bidding farewell to a great Command and its personnel. He spoke with emotion:

The time has suddenly come to say farewell to our units and stations — very suddenly, but I am glad that in spite of the short notice, I have at least this opportunity to meet you and speak to you. At the same time, you have no idea how saddened I feel by this occasion, the tremendous sense of loss that I feel on leaving this Command. I think I can without exaggeration say that my 18 months in CAC have been amongst the happiest and most satisfying of my career, happiest emotionally, and satisfying professionally. It is amongst the happiest because of the boundless affection that you have shown me throughout this period. It is a memory that I will indeed cherish. And professionally very satisfying because of the degree of progress that we have jointly made in achieving our goals and our objectives. I have been very conscious that throughout this period each of you has done his utmost, to the limit of his skill and his capacity to accomplish in his field, the tasks assigned, and to achieve the objectives that had been set. And because of your efforts, because of your work, I am able to leave with a sense of fulfillment because the record of Central Air Command is certainly something to be proud of; in the field of operations, in the field of maintenance and, no less in the field of administration.

In many, many areas much has been achieved and I take this opportunity to express to you my deep appreciation of the effort that this station has made, and the contribution of this station to the overall accomplishments of Central Air Command. Indeed, this period with Central Air Command has redoubled my confidence in the operational capability of all our fighting formations and my confidence in the capacity of all our personnel to rise to the occasion whenever and wherever the need arises.

The stations of Central Air Command, the wings, the squadrons of Central Air Command may not be close to the front, but without a shadow of doubt each will play an important role should hostilities be thrust upon us once again. Certainly, our aircrew will be across the front line in a matter of hours. And for the aircrew to accomplish their task, they will need the complete and total backing of the support echelons, the men on the ground, both officers and men, both Servicemen and civilians who man our squadrons, our units and our wings. May I stress upon you once again the vital importance, the crucial importance of ensuring that we must all constantly apply ourselves, individually and collectively to practising our operational tasks, updating our techniques and improving our professional skills. I have said to you on more than one occasion and I say again that the more you sweat in peace, the less you bleed in war. Keep this constantly in your mind and let it be your guide in all your actions during times of peace. In peace, you have the luxury of time and this is your opportunity, our opportunity, that we in the Services can neglect only at our peril.

Whilst I am naturally greatly saddened at the thought of leaving Central Air Command, this station of Bamrauli, and very specially the personnel who man it, a great source of consolation to me is the knowledge that in my successor you will have, CAC will have, an extraordinarily fine officer, a man who has not only intimate knowledge of operations but one who is known for his humanity, his sense of justice and one who will certainly give your welfare and your well-being his constant attention. I am confident that under his guidance, in his care, CAC, will continue to grow from strength to strength and continue by its actions and its accomplishments, its discipline and its dedication, by its prowess and its performance, to add lustre to our Service and to our nation.

CHAPTER 19

Maintenance Command: A Vital Link

It was 1976, when Idris took over Maintenance Command; and from Allahabad — Bamrauli to be exact — we moved to the Maintenance Command Headquarters at Nagpur. He was greatly saddened at having left a Command where he had been able to fly to his heart's content. Earlier too, he had been mostly with fighter stations. However, he found Maintenance Command a new challenge and an exposure to key elements as the base repair and equipment depots manage logistic and technical support for the whole Air Force. He soon got fully involved in improving whatever was required, as has always been his way. Idris and I soon discovered that the Air Force environment at Headquarters Maintenance Command, and interaction with officers and men and their families there, and at its many depots and all its other units, was also as heartwarming as ever.

We were allotted the official Command House, a comfortable, large old colonial type house that was earmarked for Idris' appointment. It had five bedrooms, two of which we never used, a verandah in front and a vast area all around, part of which was a laid-out garden. I loved to look after it and to be able to have a garden full of lovely flowers. Rather frightening, though, was that there were also numerous monkeys that wandered all over our tiled roof. I was really scared of their attacking me whenever I ventured out. Fortunately, there was a covered porch attached to the front of the house for the car to drive into, so we, as well as our visitors could get in and out in relative safety. The monkeys never came near anyone so no one got hurt, but, nevertheless, we kept our distance from them as they raced around on the roof and the open ground as well as up and down the trees.

The Officers Mess was a magnificent building at the Air Force base at Nagpur, and the school for the children was also of a high standard there. They were fortunate to have a very good principal at the time. Both the airmen, the officers and their families seemed happy and content at this station as it had more than adequate basic facilities for both work and recreation; and was also not far from the main town. One gets so used to the total security at an Air Force Station, that one tends to forget how precious this is compared with those in the towns. There are no burglaries, no harassing of girls, and when husbands go elsewhere on temporary duty, friends and neighbours are always within easy reach for help at short notice if ever required. A well-equipped medical centre is also available, and in most stations, there are good sports facilities, shops for fresh vegetable, fruit and other stores, and an AFFWA shop and vocational training centre too. In a very short time, I was also fully involved in all the welfare and social activities. The people whom we met in town were also very warm and hospitable.

There were Maintenance Command Equipment depots and base repair units in numerous regions all over the country. After having taken over, Idris felt it would be useful, in fact, necessary, to make visits to all his stations and units to familiarise himself with the resources and facilities of the depots/units. He wanted to gain first-hand knowledge also about the tasks entrusted to them, and the problems and difficulties that they faced, so that they could jointly examine them and seek solutions and remedies. Travelling to each with Idris was a valuable education for me as well.

Idris' address given below, at a station a year after taking over Maintenance Command, reflects his thoughts. Whilst stressing various important aspects that should always be borne in mind, he , I found, often concluded:

It is our capacity to work in peace-time and the quality of that work that will determine, without doubt, the measure of success we achieve in war — should hostilities be thrust upon us once again. Our squadrons and frontline units are undoubtedly the teeth of our Air Force, but, without doubt, also our depots and other units (of Maintenance Command) constitute its backbone. The role of each is equally vital — indeed, in peace-time, it is perhaps the most vital of all.

Your achievements over the last year have been commendable, and certainly, you and Maintenance Command as a whole can be proud of what

it has accomplished during the year that is behind us. Indeed, the country as a whole, under its dynamic leadership has made remarkable progress in many vital fields of activity. In that sense, Maintenance Command has kept in step with the rest of the country. This year, new objectives, higher targets, really ambitious goals have again been set. I am confident that this station and Maintenance Command will again prove worthy of the trust that is reposed in us.

Whilst I call upon you to make this effort, I would also like to take this opportunity to assure you that the problems that you have, the limitations under which the depots/units have to function, or the hardships that the personnel have to bear, will constantly receive my personal attention, and we will try and do our utmost to minimise them. But you, in turn, must also remember that neither I, nor any of us, have a magic wand, and solutions and remedies take time and usually money to implement. In any case, these problems, these difficulties must be seen in their proper perspective, relative to the conditions under which millions of our countrymen elsewhere work and live. Let us think less of what the country, our Air Force, or the unit can do for us; let us rather think of what we can do for our depot/unit, for our Air Force and for our country.

Maintenance Command personnel must have put in their best efforts, as the number of those who worked with great dedication and went on to receive awards went up considerably. I am adding herewith excerpts of Idris' address at a ceremony a year later where a large number of the Maintenance Command personnel received well deserved awards. In his extempore address he expressed what he sincerely felt:

It has been my proud privilege to recognise this morning some outstanding personnel of this depot/unit, some commended by me and some by the CAS himself. I congratulate them and wish them continued success. However, I would like all of you to know that I am also very conscious of the fact that whatever this depot/unit has achieved, whether the Air Force has achieved, whether the laurels that have been won on the field of battle, or the kudos that we have earned by the manner in which we have accomplished normal peace-time duties and responsibilities, all these achievements are not only

the efforts of the few who are singled out each year for awards, but the dedicated hard work of many, many more who are nearly if not equally deserving of recognition.

I would also like to take this opportunity to compliment all of you on the high standard of the parade. Let us remember that drill, certainly the movement with it, are essential elements of the deportment of Service personnel, and it is important that all of you who have obviously worked hard to achieve this commendable standard should not, with the passage of time fall into slack habits. Wherever you are, whatever you do, sloppiness will never earn you the respect of your subordinates or your superiors. Let us also remember that in times of peace the reputation that a Service enjoys depends a great deal, indeed largely, on our conduct, our deportment and our discipline not only on duty, but perhaps even more so, when off duty. Honour the uniform that you wear, be sure that you never disgrace it.

And now, turning to our work: the more I visit and see, first-hand, the departments and units of Maintenance Command, the more conscious I am of the tremendous responsibility that rests on the shoulders of all of us who man it. The resources, the equipment, that have been procured, the facilities that have been created and entrusted to our care are truly of staggering proportions, costing hundreds and hundreds of crores of rupees — procured for us at a cost that our country can ill afford. It is our responsibility, yours and mine, our joint responsibility, to ensure that we make the best possible use of it. It is a sacred trust, not only because all this valuable equipment has been entrusted to our care but also, as I keep on reiterating, because this strength and capability of our Air Force, of any organisation for that matter, depends not only on what we have but what we do with what we have. We must constantly remind ourselves that in peace-time there is no more important activity for us in this Command than to enhance output, search for new ways, better ways of logistic support, the highest possible output of our depots to build up and sustain at optimum levels the strength and serviceability of our front line squadrons. And in this, every depot of Maintenance Command, be it a repair depot or an equipment depot, has a vital role to play.

Idris went on to stress, as he had on other occasions that,

The equipment that they have acquired is expensive equipment, sophisticated equipment, up to date equipment — indeed, no developing country in the world, for sure, can boast of a production and maintenance structure such as the one that we have. The government has discharged its responsibility; the rest is up to us.

Let me tell you something else: my visits to my units, to the depots and parks of Maintenance Command have deeply reinforced my conviction, that the technical and other skills of our personnel — be they officers or men, Servicemen or civilians — in skills and in experiences, we are second to none. In ingenuity and enterprise, we are second to none. Your will, your determination, your resolve, your hard work has already achieved much; let us resolve this year to surpass the targets that we have set.

While I call upon you to make this extra effort, I want to assure you that I am also aware of the problems that many of our personnel face, problems that are especially acute at some places. Shortage of married accommodation is one; schooling for our children is another. All that I can tell you is that your Station Commander and I, and also those at Air Headquarters at the highest level, are deeply conscious of this and each day, each week, each month, all of us together are making a sustained effort to quicken the pace of construction, quicken the pace of improvement and the expansion of existing facilities.

Here again, he mentioned a point that he often stressed,

When looking at our problems, it is only right that we should also look at the problems that millions of our countrymen face. We then see our own difficulties in proper perspective. Look back also on what we were and what we are today. How we lived and worked when Partition came — and that was not so long ago. Just one single permanent station came to us, Ambala; all others went to Pakistan. Leave alone married accommodation and schools, we lived and often even worked under canvas with little or nothing to keep us occupied except hard work and the hope of a better

and brighter future, and, above all, the spirit of service before self. Today, whichever direction you turn, we have innumerable permanent stations, well-built, well-equipped and with all kinds of welfare facilities, facilities that we did not even dream of when we joined the Service.

But remember, camps and homes are not built by brick and mortar alone; each one of us has a contribution to make. As we today enjoy the fruits of the initiative and enterprise and hard work of those who have gone before us, let us resolve to play our own part in full. If each of us does that, thinking less of oneself and more of the future, genuinely and sincerely, our Air Force without a shadow of a doubt will continue to grow from strength to strength and remain the pride of our people and the nation.

Idris often told me that what he learnt of the technical resources and the exceptional expertise of the technical and logistic personnel, both officers and men, proved particularly valuable to him when, not much later — and very unexpectedly — he was posted to Delhi, as the Vice Chief of the Air Staff. An effective and successful Air Force involves all personnel of all sections. The year at Nagpur at the Headquarters of Maintenance Command, was, again, a memorable experience.

It was whilst we were at Maintenance Command that we received an unusual honour. This was an invitation for both of us from the Prime Minister for a luncheon that she was hosting in honour of the His Highness the Sultan of Brunei in Delhi. It was arranged at Hyderabad House where the Ministry for External Affairs usually entertained official guests. There were about a dozen other guests and Mrs. Gandhi was a gracious and warm hostess, while the young Sultan was dignified but friendly, and he made a special effort to greet and have a word with each guest.

CHAPTER 20 Vice Chief of the Indian Air Force

In 1977, the Air Chief was going abroad, and Idris as the senior-most officer after him, was appointed as the Vice Chief of the Air Staff. We moved to Delhi and went to live at No. 25 Safdar Jung Road which was our home until August 1978. Idris was to have retired in June 1978, but his service had, some months earlier, been extended by the government by three months. In May 1978, while Idris was on a two-day visit to Hyderabad, I received a call from the Defence Secretary Sushital Bannerji asking for Idris, as he had something of very special importance to announce to him and he needed to personally speak with Idris right away. By a happy coincidence, Idris was with his mother when Sushital then informed him that the government had just passed orders that Idris should be the next Chief of the Air Staff of our Air Force. It was so, so wonderful, especially also that Idris was with his loving and proud mother when he received the news. God has truly been so kind.

Idris was naturally elated. When I mentioned what a great responsibility he would have to shoulder, he said that undoubtedly when hard decisions had to be made, finally, they would have to be his. But what always greatly helped him when taking a final view was what Sarojini Naidu, who was a friend of his family and mine, had stressed to him when, as a 10-year-old, he, with his parents had visited her: "Above all, *obey the dictates of your conscience,* God's great gift to each and every one of us." "If one's conscience is clear," Idris added, "it gives one not only great strength but, no less important, great peace of mind."

Idris and Bilkees with their children, left to right: Mariam, Asgar and Asad in Delhi at 25 Safdar Jung Rd, 1977

It was in this context, as an example, that Idris then went on to tell me something which he had never even mentioned to me earlier. This was about a new challenge that had to be faced soon after, but, Idris added that, however, complex the problem, whenever in doubt, his conscience was his guide. On that occasion, Idris said that one morning, soon after getting to his office, he received a message from the Ministry of Defence that the Defence Minister, Shri Jagjivan Ram would like to see him. Normally, a message of this nature would have come through the Air Chief, but at that time ACM Mulgaonkar was out of town. Shri Arora, the Personal Assistant to the Defence Minister had himself spoken to Idris. Shri Jagjivan Ram, a person whom

Idris greatly admired and respected, and whom Idris had earlier met on several occasions, met him very cordially, asked him to sit down and offered him a cup of tea. Moments later, however, the shock came: This was about a serious matter regarding Army Aviation. "I have something very important to convey to you, Air Marshal," the Defence Minister told Idris. "It is well known that you have very strong views about the continued command and control of the AOP Flights (Air Observation Posts) of the Air Force, by the Air Force, and are, therefore, strongly against their transfer to the Army, as is being constantly pressed by them. I felt that it is important that I should mention to you that government is accepting the Army's recommendation."

After a minute's silence, Idris told me, long enough to recall what Mrs. Naidu had told him — and which had become his guiding principle ever since — he told the Minister that to be true to the country, which had placed its trust in him and also to him as our nation's highly respected Defence Minister, he would be absolutely honest. He then added that he would, therefore, have to tell him that if the change was to be implemented when (as he had already been informed) he was appointed as the next Chief, his conscience would not allow him to accept this decision because he was genuinely convinced that it could well prove a disaster should any war be thrust upon the country once again. Consequently, to be true to the trust that was being placed on him, if this came through while he was the Air Chief, he would feel compelled to resign.

Idris repeatedly told me that he has, even now, no doubt in his mind that it *was, genuinely,* his conscience that largely determined his decision. Knowing that this would be a shock to the Defence Minister, a person who, Idris said, had always been exceedingly considerate with him, he went on to say that since he (the Minister) always encouraged others to speak freely and frankly, he wanted to make two or three submissions: firstly, he went on to say, as is well-known, across the world, every country has a separate Army, Navy and Air Force. It is a recognition of the fact that true optimisation of each necessitates a lifetime of specialisation; and *the expertise and employment of air power is no different.* The rapid advance of science and technology, Idris added, has only increased its importance. Idris mentioned to me that he took special care to emphasise that he had nothing against the Army, and this was no inter-Service rivalry; adding that, in fact, throughout his service he, with the rest of the nation, admired the

professional skills of the Army in their field of specialisation. Also, that he had very close and very warm relations not only with its Chief but, over the years, with many senior and junior Army officers alike. Noteworthy also was the fact that after every conflict since Independence, the Air Force had received heartwarming letters of appreciation from its sister Service at the highest level, conveying the Army's deep appreciation and gratitude for the Air Force's exemplary support.

Finally, his own knowledge and his nearly 40 years of service had convinced him and others in the Air Force that in war, on the battlefield, coordination *and control of the employment* of air power, often based on different bases, and ensuring effective close support to the ground forces, particularly in swiftly moving battles on the ground, *unified command and control* of air power by the Air Force was *vitally important*. If part of it (i.e., the AOP Flights) was to be trained and operated by another Service, as was being proposed, it could well prove disastrous. This was in a way, proved by the Vietnam War, where divided control of air power in support of the land forces was a major factor in the prohibitive losses of battlefield support aircraft and, in all probability, the ultimately decisive reverses suffered by the West.

Keeping all these factors in mind, Idris told the Defence Minister that the proposed change was most inadvisable. An alternative, Idris finally added, could be for the Army to be asked to indicate the areas in which they have experienced problems in recent wars, and the Air Force could be first given an opportunity to develop alternative systems and procedures to overcome them. To Idris' relief Mr. Jagjivan Ram smiled, indicating, perhaps, that he accepted Idris' submission. Smilingly, Idris concluded to me, the AOP Flights thereafter remained with the Air Force.

I recently asked the very distinguished Director General of the Centre for Air Power Studies in Delhi, Padma Bhushan awardee Air Commodore Jasjit Singh if he could let me know of anything interesting for this book about Idris. I am most grateful to him for taking the trouble to write back. I received a gem of a letter from him, some excerpts of which I am herewith adding:

> *I recall my days in Air HQ in late 1977 as a Wing Commander, Deputy Director Flight Safety (Prevention), a job I took rather seriously in spite of what had happened a couple of months earlier. I worked hard, even if I say so now 35 years later, because I was perhaps trying to prove a*

point — to myself. And I have many memories of that period. One relates to the reactions of the Vice Chief of the Air Staff (then Air Marshal) Air Chief Marshal Idris Latif, later CAS.

You see, the Vice Chief of the Air Staff Air Marshal Latif had given a blank cheque to encourage every officer in his branch to feel free to write directly to him on any matter which in the author's opinion would help to improve the Service and quality of life in the Air Force. The only limit was that the paper should not exceed one page! After I had been in the job for about a month, I decided to write a note to the VCAS. Understandably, my friends and colleagues tried to discourage me. Somebody said the VCAS has more important things to do than read your silly note! This only strengthened my belief that if the VCAS had given the junior officers like me such privilege, then they should come up to justify his faith, especially since the guideline clearly was that the subject matter must focus on the betterment of the Air Force.

The Air Chief had always been extremely conscious of the critical need and value of high morale of air warriors — and in particular of the 700-odd fighter pilots who would be right in front at the cutting edge of combat. Our air warriors were the real strength which had defeated the high technology Pakistan Air Force in 1965. This, he felt, was as much due to the faith and dedication of all personnel as it was of the top leadership. As Air Vice Marshal, he had headed the official study team just after the 1971 War, and visited all the fighting units to find how true was Winston Churchill's famous tribute to the aircrew after the Battle of Britain in 1941, when he said that never before in the history of human conflict so many had owed so much to so few!

In a large organisation like the IAF, he knew very well that an odd officer would prefer to lead through coercive authority. But this was a rarity, an aberration that Latif was still willing to forgive in the hope that this might re-orient the errant leader. After all, the true leader is one whom those who are led would want to lead them, and they could follow with dignity and pride in the Service.

CHAPTER

21 In the Hot Seat

Idris became Chief of the Air Staff of our Air Force on 1st September 1978. As he often said, three years is not a long time to produce meaningful results, but the modernisation of our weapon systems was imperative. Deep-strike aircraft, fighters, reconnaissance aircraft, transport aircraft, combat helicopters, many, if not most, had become outdated if not obsolete. At the top of the list was a deep-strike aircraft to replace the Canberras. This had concerned him, as already mentioned in an earlier chapter, since his time as Assistant Chief of the Air Staff (Plans) in 1971-73. But apart from modernisation, another matter of equally, if not greater concern to Idris when he would be taking over as Chief, was that, at least in his view, measures towards the upkeep of the morale of the aircrew and field commanders were no less important.

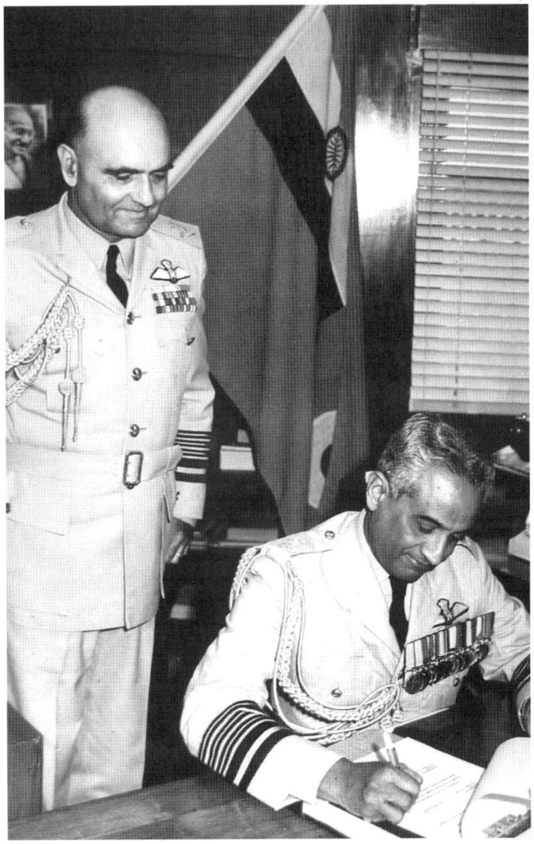

Idris taking over as Chief of the Air Staff, w.e.f. 1st September 1978

Idris calling on Shri Jagjivan Ram

As Idris saw it, the current methods to reduce flying accidents were affecting the combat readiness of the force. An added factor was the dismal career prospects of those batches of officers who had been recruited during the period of the Chinese aggression of 1962, who had now reached the selection for promotion grade (as distinct from 'time' promotion). Better career prospects, especially for these ranks, particularly for the large numbers that were then, coincidentally, manning the cutting edge of the Air Force, was not an easy task, particularly with a tight-fisted Ministry of Finance! To make matters worse, for the non-commissioned ranks too, the prospects had become unacceptable. For example, Idris once pointed out, it took 18 to 19 years for an airman to become a Sergeant in certain trades, against 12 to 13 in others. This, he strongly felt, must be remedied. These were some of his foremost concerns, Idris said to me, when, on 1st September 1987, he walked across from his Vice Chief's office to the Chief's, barely twenty yards down the corridor.

Idris calls on President Sanjiva Reddy after taking over as Chief of the Air Force

I was reading Idris' speeches and other notes in his old files when I came across an interesting note on his tenure as the Chief of our Indian Air Force. Updated excerpts of Idris' notes are given below:

Idris as Air Chief reviewing the Air Force Anniversary Parade

MY THREE YEARS AS CHIEF OF THE AIR STAFF

On 1st September 1979, as I stepped into the spacious office of the CAS, I should perhaps have felt somewhat awed and daunted by the tasks that lay ahead. Instead, I clearly recall, I felt cool, collected and confident. Partly, I am sure, it was so because I was very clear in my mind of the goals that I had set for myself and the initiatives that I felt were necessary to achieve those objectives. Equally important, I knew that I was fortunate in having with me, to form my team of senior field commanders and Principal Staff Officers, some of the finest men one could wish for, officers who were not only professionally outstanding, but officers whose loyalty and integrity I could totally rely on. I selected Air Marshal Dhatigara as my Vice Chief of the Air Staff and Air Marshal Dilbag Singh as the Commander-in-Chief of Western Air Command with Air Marshal Katre as his Senior Air Staff

officer. Air Marshal Grewal I put in charge of the Eastern Theatre, and Air Marshal Chhabra as the Commander-in-Chief of Training Command. To head the Maintenance Branch, I had Air Marshal Naik. As my Staff Officer, I selected (then) Wing Commander Krishnaswamy of the Education Branch, an officer who had that rare combination of qualities of exceptional ability and heartwarming humility. With such an outstanding team, little wonder that I was able to assume my responsibilities with a feeling of complete confidence, impatient only to push forward with the initiatives that I strongly felt were both necessary and urgent.

In my first address to the Air Officers Commanding-in-Chief (AOC-in Cs) and Principal Staff Officers (PSOs), I first spoke at length on the subject of human relations. To them and to all my other Commanders, I said that morale is recognised the world over as a key element in any fighting force, but there was no doubt in my mind, I said, that fear of punishment could never be a satisfactory basis for running a Defence Service — indeed, any organisation. It was my firm belief that human resources could be optimised only on the basis of trust, confidence and true loyalty. I recognised, I added, that though fear, unfortunately, cannot be totally eliminated, it must be the exception rather than the rule, and the central thrust of man-management must be structured around the more positive and constructive elements of human relations. My faith and trust were rewarded by events: even the flying accident rate in the succeeding months showed a marked decline.

While I was absolutely sure my Commanders would faithfully implement my directions, demoralisation, I knew, had become so distressing, and, indeed, damaging to our operational capability, that I felt I should visit at least all the major stations as early as I could, not only to establish a personal contact at key levels but also to underline personally the changes in man-management that I felt were so vitally important if our Air Force was to remain an effective fighting force. On these visits, therefore, first and foremost, I stressed that we must jointly ensure that our Air Force was held together by bonds of true comradeship, bonds of trust and loyalty and an identity of objectives. I was fully aware that apart from morale, the

other most vital area that needed urgent attention was the replacement of some of our obsolescent weapon systems, some of which, in fact, were totally obsolete. In spite of this, I felt that it was necessary to remind all personnel that it should never be forgotten that the equipment and the facilities that had been procured for our Air Force and placed at our disposal had cost the nation thousands of crores of rupees — procured for us at a cost that the country could ill afford.Therefore, whilst modernisation of our frontline aircraft , particularly the strike element, was amongst my foremost concerns, it was our responsibility, I said, theirs and mine, our joint responsibility, to ensure that we made the best possible use of existing aircraft and other weapon systems.

The threat to our country was real, and not only Air Headquarters, but also the government at the highest levels, I assured them, was fully aware that in terms of weaponry we enjoyed no decisive advantage and that several systems needed to be updated. But, I emphasised, the long history of war had proved, if nothing else, the supreme importance of the human element in the total weapon system. The outcome of any future conflict, I reminded them, would also substantially depend on our planning and our training and our ingenuity. In the ultimate analysis, I said, it would be our capacity to work in peace, and the quality of that work that would determine the degree of success we achieved. This outlook, I said, must become central to all our endeavours.

The importance of devoting time and attention to the welfare of personnel under one's command, I went on to say, was well-known and recognised. But in determining the initiatives, we must not forget that contentment was essentially a state of mind and not dependent solely, or even largely, on material comforts. In our efforts to improve the working environment, a close and healthy personal relationship with one's subordinates was of prime importance. Another aspect, often lost sight of, was the need to make personnel constantly aware of the many benefits they enjoyed, especially when measured against the needs of the vast majority of our countrymen. When we look at our problems, I said, we should also find time to ponder over the problems that millions of our countrymen faced; our own difficulties and wants would then be seen in their proper perspective. In any case, let us

not forget, I added, how we lived and worked in yesteryears. When Partition came, Ambala was perhaps the only permanent station that came to us, all others went to Pakistan. But today, whichever direction one turned, from Srinagar in the north to Bangalore in the south and from Chabua in the east to Jamnagar in the west, we had innumerable permanent stations, well-built and well-equipped with a host of welfare facilities we did not even dream of when we joined the Air Force.

Considering that I am to cover three full years of my tenure as CAS in a few pages, it would seem that I have given disproportionate space to the sentiments I conveyed to the officers and men during those first few weeks. But I have done so deliberately because those were central to the actions and initiatives I took during the years I was to preside over the fortunes of our Air Force.

On my return to Delhi, I gave top priority to the modernisation of our offensive weapon systems. In particular, I was greatly concerned with the very serious limitations of our Canberras, our only deep-strike force. Government had, for several years accepted, in principle, the need for their replacement and, after very comprehensive evaluations, all three of my predecessors had strongly recommended the early induction of the twin-engined Anglo-French Jaguar initially selected when ACM Lal was our Chief. I was fortunate to be under him as the first ACAS (Plans) with overall responsibility for assessing its suitability as the low-level deep-strike aircraft to succeed the Canberra. And, happily, for this, we all strongly favoured the Jaguar over the two other contenders, the French Mirage FI and the Swedish Viggen. The Jaguar, unlike the other two, was not only twin-engined (particularly important, we all felt, for survivability in deep penetration at low levels, especially in our environment); also, because its relatively high bypass engines gave it a far greater radius of action at low levels. And, finally, of very special importance, it had a tailored low-level navigation and attack system.

ACM Lal, a brilliant and much respected Chief, held in very high regard not only in the Defence Services but also at all levels in government, did

his best for its early induction. But, unfortunately, though approved in principle, the Air Force was told year after year after year, that paucity of funds and other national needs, would not permit its early procurement. For each succeeding Chief, we all knew, it also remained a very high priority task, and each, without doubt, did his best to convince the government of the extreme importance and urgency of its induction. But again, the government, though never questioning its importance nor the urgency, primarily because of extreme paucity of funds, had no option but to defer the acquisition. When I was appointed CAS, the Janata Party was in power, with Mr. Morarji Desai as the Prime Minister. In my effort to force the pace of the induction of the selected Deep Penetration Strike Aircraft at the highest level, I invited him to Air Headquarters to present to him, in our Operations Room, my analysis of the threat, and the strength and limitations of our Air Force. Happily, he immediately agreed. With the Prime Minister came, as I recall, Shri Charan Singh (then Home Minister), and Shri Subramaniam, then Defence Minister. Accompanying them was the Defence Secretary, Shri Shushital Banerjee, an outstanding IAS officer, and always a pillar of strength to me.

I decided on a short and simple presentation, concentrated on a graphic comparison of our strike capability with that of our immediate potential adversary, the main thrust being on the relative 'throw-weights' and radii of action of the strike aircraft of the two Air Forces involved. After the briefing, escorting him to the lift, he looked at me with his rather charming smile and said to me in Hindi, "Mujhe darate ho?" (Are you trying to frighten me?) "If the facts are frightening, Sir", I replied, "remedial measures are clearly overdue". Again he smiled; and, to his great credit, — and the Air Force's undying gratitude — within the week he called a meeting at his residence, at which, after barely a few words, he directed me and Sushital Banerjee the then Defence Secretary, accompanied by two or three (I can't recall exactly) senior officers — I think they were the Finance and Defence Production Secretaries — to proceed immediately to the UK to negotiate and finalise the procurement of four squadrons of the Jaguar — all that we had proposed.

In London, very high-level meetings were promptly arranged by our High Commission at which we (Shushital Banerjee was the main spokesperson) not only strongly underlined the importance of the cost factor, but also, as earlier, most strongly urged by me, the extreme urgency of our requirement. The British were clearly impressed, and, to our relief, not only did we find that their final quotations were competitive, but, unbelievably, to my delight, to accelerate deliveries, they finally also agreed to temporarily run down and 'number-plate' one of their Jaguar squadrons and make available to us on loan and at no additional cost (!), a full squadron's complement of aircraft. This was to tide over the period till deliveries against our own orders would begin to flow in. As a direct consequence, the first Jaguar touched down in India within about three months of signing of the contract.

On our return, I decided on Ambala as the base for the Jaguars, then commanded, as I recall, by Air Commodore Bawa, and 14 Squadron under the command of Wing Commander Nadkarni, one of our outstanding fighter pilots, as our very first Jaguar Squadron. The Personnel Branch was then asked to select 11 more of our outstanding fighter pilots, to be sent, under the overall command of Wing Commander Nadkarni, to England in batches of four, for conversion on the Jaguar as had been earlier arranged with the RAF. This was all done primarily to get our new deep-strike force operational as quickly as possible. I also nominated Wing Commander Shashi Ramdas, an outstanding technical officer who had earlier done splendid work under me at No.2 Wing, to be the Senior Engineering Officer of the wing. Fortunately, being already cleared for promotion, he took over in the rank of Group Captain. These decisions, I have to say, I never regretted. At this stage, I cannot help but also mention how very proud we all were (and are) of the laurels our carefully selected pilots earned for our Air Force by their exceptional performance. In fact, Air Marshal 'Tiny' Kumaria, our VCAS today, if I remember right, broke the gunnery record by a score of 100! And imagine, to crown it all, of the twelve we had originally selected, by now, ACM Tyagi and ACM Browne, two great Chiefs, and two others, Air Marshal McMahon, and Air Marshal Kumaria, no less outstanding Vice Chiefs. It was a matter of deep personal satisfaction to me also that, before I retired,

our low-level deep-strike force of Jaguars, though still limited in size, was operational.

The Canberras were not the only aircraft that needed replacement. The Hunter, though one of the prized aircraft on our inventory during the Sixties and early Seventies, was ageing, and the MiG-21 and the Sukhoi-7 had too limited a radius of action in the strike role. It was again our good fortune that during a very high-level visit of the Soviet Defence Ministry, their Air Force Chief who called on me at Air HQ during our discussions, most unexpectedly, offered to make available to us as soon as we felt we could absorb them, a variety of their frontline aircraft. I could hardly believe my ears, particularly when he included, specifically, the then highly classified MiG-25. I strongly felt that we should not lose such a golden opportunity to modernise our Air Force, particularly at the extraordinarily low Soviet prices and almost unbelievable credit terms. Considering both the magnitude and urgency, I again, with the approval of the Defence Minister whom I fully briefed, requested a meeting with our Prime Minister, who was Shrimati Gandhi. Briefly, but with care, I explained to her the importance and urgency of our need for the aircraft being offered, stressing also the very great financial advantage of the Soviet prices and exceptional credit terms. After some thought and a brief discussion, she said, to my delight, that she favoured the idea and asked for specific proposals.

Of course, at Air HQ, we were fully aware of the dangers of taking on too much too quickly, particularly from the maintenance point of view. After the most careful consideration and evaluation, including flight tests by our test pilots, we proposed, and government immediately accepted, a relatively small unit of the reconnaissance version of the MiG-25, a more substantial number of the MiG-23 to augment our ground attack capacity, a number of suitably modified An-32s, plus some attack helicopters. The replacement of the An-12 fleet, we felt, could safely be deferred. Before I retired in 1981, in addition to the Jaguars, the MiG-25s and MiG-23s and, if I remember right, a few modified An-32s also, were in our frontline squadrons. In fact, on my farewell visit to Mumbai, I was happy to learn that most of the new helicopters had also arrived and a few were already assembled!

I must recount another rather memorable event. During the visit of the Soviet delegation, the Defence Minister mentioned to me that he had invited our government's participation in their forthcoming manned space satellite launch by including an Indian officer, preferably from the Air Force. Thrilled, I assured him that he could count on my wholehearted support. Keen particularly to add this feather to the Air Force cap, that very afternoon I sought a meeting with the Prime Minister, Shrimati Indira Gandhi. As always, she was very receptive, but added that, for some reason, the head of ISRO, Mr Dhawan, was not in favour and she was reluctant to overrule him. She then asked me if I would like to speak with him. I told her that I knew Mr. Dhawan well, and I would be delighted to do so. The very next morning I spent almost an hour with Mr. Dhawan. The main thrust of my submission to him was that whereas we were all aware that in our country's space programme, manned flight into space was only a distant dream, the first-hand knowledge that we would gain — and at no cost — would, without doubt, I felt, prove invaluable. Our very cordial relationship helped, I think, and I was delighted when he finally said that, whereas it would be difficult for him to support me, he would not oppose it. Not wanting to waste time and risk the possibility of any fresh complication, I requested 'just a few minutes' again with Mrs. Gandhi that very afternoon and, when I told her of Mr. Dhawan's response, I was thrilled to see how pleased she was.

Two Air Force officers were carefully selected, a Wing Commander Ravish Malhotra and Squadron Leader Rakesh Sharma. Months later, both underwent intensive training in the Soviet Union, the latter being finally selected by the Soviet authorities themselves. Finally, not long thereafter, Mrs. Gandhi from her office, spoke to Squadron Leader Sharma in outer space: "Squadron Leader Sharma, what does India look like from up in space?" And Sharma replied, "Madam, Sare Jehan Se Achcha, Hindustan Hamara"! Apart from the priceless knowledge and first-hand experience, another memorable feather in the Air Force cap!

I cannot help but mention that two fine civil servants, Dr. Alexander and Rao Sahib Krishnaswamy, both, in turn, Principal Secretaries to the Prime Minister, were always of great assistance to me in my efforts to obtain government

approvals, including those for two other important measures in my final months as Chief, namely, a very major increase in the higher vacancies for our non-commissioned ranks, a programme that made the relative career prospects of their respective specialisations far more equitable; and, secondly, a generous scale of rations for our aircrew. I was able to implement another matter that had concerned me. Limited but carefully selected free rations for aircrew also, I strongly felt, were very desirable, particularly as doctors had advised that, as a matter of flight safety, a good breakfast before flying was absolutely necessary. In our Air Force, flying normally commenced very early in the morning, at hours which put a great strain on most households. Consequently, a wholesome nourishing breakfast, I felt, should be made available in the aircrew room of each squadron (I had actually been able to implement this in a very limited way with non-public funds supported by the resources of our Officers Mess when I was commanding 2 Wing in Pune in 1970). Both these measures greatly helped, I later learnt, to instill a high degree of confidence amongst all ranks that their welfare was of constant concern to me and my team at Air HQ.

Another major decision, but this time concerning our long-term reequipment plans, was taken during the final months of my tenure. In spite of the limitations of Hindustan Aeronautics, underlined by the quite serious limitations of the Marut, and also, in a few important respects, the inadequate improvements to the Gnat, we felt that the country should not indefinitely remain totally dependent on foreign powers for so important an element of national security, particularly with a large, and well established aircraft design, development and manufacturing facility like HAL available to the country. National prestige was also involved.

Keeping in mind, however, HAL's limitations, as also the continuing paucity of financial resources, I was clear that we should aim for only a relatively simple and inexpensive fighter aircraft, but one which, nevertheless, could still fulfill an important operational role. This was discussed at length and we came to the conclusion that in battlefield support of the Army, which we all considered a task of vital importance, often there were many vitally important targets such as tanks, artillery concentrations for which the use

of highly sophisticated and expensive aircraft was not really necessary — or even, in a sense, cost-effective. Such targets, we felt, a relatively simple aircraft without highly sophisticated and expensive avionics, could easily but effectively engage, provided it had adequate firepower and a reasonable radius of action and endurance. High speed and good manoeuvrability, primarily to ensure adequate survivability when returning, would also, we all agreed, be essential. A related important factor that weighed with us was also, of course, that, with relatively much lower costs that all this would result in, the number of our fighter squadrons, always a matter of concern, could, perhaps, in due course, be substantially increased. At the very least, force levels could be maintained when, sooner rather than later, we would be compelled to begin phasing out the MiG-21 and the Sukhoi-7.

The result of all these considerations was the Light Combat Aircraft (LCA), essentially also, in my view, a Low Cost Aircraft (again LCA!). After extensive discussions at which the Scientific Adviser to the government was always present, at our request, HAL, at the level of the Chairman, assured us that the relatively small and unsophisticated LCA that was discussed, could definitely be made available to us to enter squadron service by the mid-1990s. We felt that even though 15 years seemed rather long for the relatively simple fighter we had agreed upon, it would give us a reasonable degree of surety. We at Air Headquarters, however, I clearly remember, did express serious doubts about the adequacy of their estimate of Rs. 500 crore as total development costs. It is most unfortunate (in my view) that, in later years, without (in my view) the limitations of our national capability being given adequate weightage, the concept concerning performance and sophistication underwent major changes. Largely, as a consequence (again, in my view), over three decades later, the LCA is still to enter squadron service.

During my tenure, another very important matter that caused me very great concern, indeed anguish, was the steep deterioration in the career prospects of our aircrew coming into the selection grade. It was really tragic to see some fighter pilots who, even after having served for many years with great distinction, including several who were well above the average officers, being superseded as there were far less high vacancies available.

This was devastating for the concerned officers, and, understandably, no less devastating for their families, and also for many of us who had known them personally and often worked with them closely. Apart from the personal aspect, I was deeply concerned because clearly it was very adversely affecting the morale of what was really the highly skilled cutting edge of the Air Force. The cause was historic. In the mid-1960s, following the Chinese aggression and government's plans to raise more squadrons, large numbers of officers were inducted. In later years, including the early years when I was Chief, I remember, very many more than usual, became due for promotion to the then first selection rank of Wing Commander. But since, mainly due to paucity of resources, the planned expansion of the Air Force had not taken place, the selection rank vacancies had remained largely static. As a consequence, generally, whereas in normal times at least one of every four got his rank, the ratio for promotion had dropped to one in around twenty; sometimes, it was even worse. Deep selection, as it came to be called, resulted in many very able officers getting superseded; not a few, as I have earlier indicated, who had been assessed well above the average, and given command of squadrons.

It was tragic beyond words, and however much one attempted to rationalise and publicise the cause within the Air Force, it was not difficult to understand the shock, frustration and agony of the affected officers and their families. The Service as a whole suffered too: apart from the most distressing lowering of morale of a key element of our force, an understandable feeling of humiliation had driven some to even put in their papers; and highly trained manpower was lost. Their anguish was ours too, often made bitter by what many felt was the inadequate appreciation, at important levels in the concerned Ministries, of the dimensions and consequences of the problem the Air Force faced. To complicate matters further, government linked our proposals with the other two Services. The vacancies that were ultimately approved made some difference but, as expected, in the process, created fresh problems, but we had to live with them. I distinctly recall a feeling that, somehow, if we are to optimise national resources, a better and more integrated system of decision-making would have to be evolved.

Idris' note above does not include the aircraft that he so loved to fly. As for Idris' love for flying, he fully qualified himself and flew no less than 22 different types of aircraft which included fighters, bombers and even heavy transport aircraft. Even in the final years of his service as the Chief, he found time to fully qualify himself on the Sukhoi-7, the MiG -21 and the Jaguar; plus familiarisation sorties on the Mirage 2000 (during an official visit to France), and, finally, the MiG-23 and even the MiG-25 just months after the aircraft first arrived and were assembled, which was just before his retirement.

List of aircraft flown by Air Chief Marshal Latif (not in chronological order)!

Tiger Moth
Wapiti
Audax/Hart
Proctor
Hurricane
Spitfire
Devon
Vampire
HF-24 Marut
Canberra
Super Constellation
Otter
Northrop F-5 (in USA when Air Attaché in early 1960's)
Avro—748
Dakota
C-119 Fairchild Packet
MiG-21*
Sukhoi-7*
Jaguar*
Harvard
(Miles) Master II

Tempest

*Note: Conversion and some solo flying whilst CAS

Familiarisation flights only, whilst CAS

Mirage 2000

MiG-23

MiG-25

The photographs here and below are of aircraft flown while Idris was the Air Chief

MiG-21

Sukhoi-7

Idris after flying the Sukhoi-7 at Hindon with the Station Commander Air Commodore Sikand and station personnel

MiG-23

Idris' first solo in the Jaguar, 1980

Jaguar

MiG-25

Idris' note concludes:

There were frustrations also, frustrations and anxieties and disappointments; but, sooner or later, we all learn to accept them as part of life. Also, without doubt, what gave me enormous strength and deep satisfaction, I am sure, was that with the lows, hand in hand, went the deep satisfaction of having succeeded in achieving, in a large measure, the important goals I had set for myself. These included in particular, the restoration of the morale of key elements of our Air Force and, of course the major strides we had been able to take in modernising and strengthening vitally important elements of our weaponry. Coupled with these, of course, for me, was the thrill of life in our Air Force. This had been, in a sense, living and working, as part of a large family, with men whose courage and competence, whose dedication and devotion and whose loyalty and integrity was — and I know, remains — second to none. The Air Force environment had become so much a part of one's life, that for me, it was almost inconceivable that sooner or later this too, like all else in life, must end. And sure enough, one fine

day, 1st September 1981 to be exact, when I woke up and went to my dressing room, my wife had put out for me a safari suit instead of my uniform.

Air Chief Marshal Idris Latif receiving Prime Minister Indira Gandhi on her first visit to Air Headquarters (Vayu Bhavan) after his appointment as Chief of the Air Staff

Idris showing Bilkees the MiG-21, 1979

CHAPTER 22

Memory Holds the Door

In addition to Idris' note as above, there are other instances that I recall which include Idris having discussed establishing a Military Housing Board project for retired officers with the Army Chief, who also thought it was a good idea. Together, they went and discussed it with the Defence Minister who also approved, and gave a note for the project to be worked on. Unfortunately, the wording in the note stated that it was for military housing, and in no time at all, the word "military" was taken to mean the Army, and it was worked out and accepted as such. Idris was deeply disappointed and went to see the Naval Chief and discussed the creation of an Air

Idris as Air Chief with Defence Minister C. Subramaniam on his right and the Scientific Adviser Dr. Raja Ramanna on his left

Force and Navy Housing Project which the Naval Chief immediately agreed to. After it was sanctioned, with the Navy's concurrence, Idris put Air Vice Marshal Kishen Kapur, a retired Air Force officer, well known for his competence and dynamism, in charge of the project. It was started in Noida, but soon became very popular and has established branches across the country. Kishen, incidentally, later went on to become an able and popular Lieutenant Governor of Delhi.

Naval Chief Admiral R. Pereira, Air Chief Marshal I.H. Latif, Chairman Chiefs of Staff Committee, Army Chief General Krishna Rao, 1981

Idris and I with the King of Bhutan, 1980

Idris with Mrs. Sharda Mukerjee, 1980

Air Vice Marshal Krishnaswamy (Retd.) (then a Wing Commander) was Idris' excellent choice for his Staff Officer. His calibre can be gauged from the fact that he was requested to continue as Staff Officer by the next two Air Chiefs also. I am most grateful to him as, at my request, he has sent me his views and reminiscences on the tenure of Idris as the Air Chief, a synopsis of which is given below:

Appointment of Air Chief Marshal I.H. Latif as Chief of the Air Staff

The political scenario at the time of his appointment was somewhat fluid. Following the defeat of Indira Gandhi in the elections in 1977, the Janata Party had formed the government under the Prime Ministership of Morarji Desai and later Charan Singh. In 1980, Indira Gandhi returned to power. During his tenure, therefore, Air Chief Marshal Latif had the distinction of functioning under two different political leaderships opposed to each other. It was to his credit that he had an excellent rapport with both and with the respective Defence Ministers, Jagjivan Ram, C. Subramaniam and later Indira Gandhi, who also held the defence portfolio.

Role in the IAF Modernisation Programme

Traditionally, over the years in the past, national defence policy and postures had remained more reactive than proactive. The annual defence budgets as a percentage of the GDP had also been much lower than the comparative figures obtained among our potential adversaries. This inevitably reflected in the rather tardy progress of much-needed upgradation and reequipment of the Air Force. The last major upgradation programme was the induction of the MiG-21 series in the early 1960s, ushering in the era of supersonic fighters in the IAF. With the vast experience that he had gained in an earlier appointment as the Assistant Chief of the Air Staff (Plans), Air Chief Marshal Latif was deeply conscious of the need to get the modernisation programme on the rails. Keeping this in view, he arranged for a thorough examination of all available records and a reappraisal of all such matters awaiting government decision. This was necessary for making a realistic assessment of the operational capabilities of the Air Force in relation to the threat perspective and proposals still pending.

The offensive capability of the Air Force is generally represented by the quantum and quality of the strike aircraft fleet. Therefore, the focus of attention had to

be on the modernisation of this strike element. The old faithful Hunters and Canberras which for decades had carried the burden of strategic, interdiction and air defence roles were, very worryingly, obsolete. Air Chief Marshal Latif firmly believed that the priority had to be on upgrading the strike capabilities of the Air Force. The requirement was for a Deep Penetration Strike Aircraft (DPSA) for which the Anglo-French Jaguar had been selected almost ten years earlier and approved in principle by the government, when Air Chief Marshal Lal was the Chief (and, subsequently, by two of his successors). But this could not be acquired by the government primarily, we were given to understand, because of paucity of funds.

Realising that the best approach was to bring the imperatives of modernisation to the notice of the government at the highest level, ACM Latif sought and was given an opportunity to make a presentation to the Cabinet on the current status of the Air Force. The full-fledged Cabinet Committee of the Janata government, including the Prime Minister Morarji Desai, Charan Singh, Jagjivan Ram and C. Subramaniam, was present at the operational briefing at Air Headquarters. With well-documented facts and figures and a comprehensively illustrated map, Air Chief Marshal Latif brought out the operational status of the Air Force, its strengths and deficiencies in strike capabilities and the gaps in the air defence environment. He also highlighted the delay in finalising and approving long pending acquisition programmes, in particular, the induction of DPSA. His frank exposition was well appreciated and obviously made a strong impact on the Ministers, so much so, in fact, that Shri Charan Singh remarked "Air Force ko kya chahiye" or words to that effect. Further, in a one-to-one interaction with the Chief after the presentation, Morarji Desai enquired about the DPSA and was informed of the options and the preferred choice of the Air Force since Air Chief Marshal Lal was the Chief. The Prime Minister responded immediately and directed that between the Chief and the Defence Secretary Sushital Banerji and the Secretaries of Finance, and Defence Production, immediate steps should be taken to finalise the deal for the acquisition and induction of the British Jaguar.

The discussions with the RAF were particularly useful as they paved the way for free and full exchange of information on the operational, logistics and financial aspects of the Jaguar aircraft. The broad outlines of the progressive

manufacture of the aircraft at HAL also emerged from the discussions. In view of the great urgency of our requirement, repeatedly underlined during the discussions, the RAF went out of the way to even number plate one of its Jaguar squadrons and divert the aircraft to the IAF. It also undertook to train several batches of IAF pilots as well as engineers and airmen on the Jaguar.

Immediately on ACM Latif's return, Air Headquarters formed the Jaguar Project Team (JPT) to be located at Wharton to work out the details of the induction of the aircraft into the IAF. The then Group Captain (later Air Marshal) B.D. Jayal, (later Air Marshal) Ramamurthy as the technical expert and (later Air Vice Marshal) Kasinath as the logistics expert formed the team. Air Chief Marshal Latif personally briefed the team, highlighting the need for a thorough and close scrutiny of all aspects so as to ensure smooth and early induction of the aircraft on terms most beneficial to the IAF. The team even scrutinised with great care the cost of individual items of spares, identified some of them that appeared to be unduly highly priced and was able to either bring down the cost or identify indigenous substitutes.

The first Jaguar squadron of the IAF was made operational in September 1980, and the Chief proudly showed the new acquisition to the then President Neelam Sanjeeva Reddy. The next squadron came into being a year later and others followed in quick succession.

Indigenisation

Air Chief Marshal Latif maintained a close watch on indigenisation efforts in reequipping the IAF. However, while being an ardent supporter of indigenisation, he was also deeply concerned about the frequent overruns of accepted time-frames for development and delivery of proven systems, and insisted on frequent reviews with DRDO and HAL at high levels.

Welfare of, and Loyalty to, Subordinates

His soft-spoken and pleasant personality, combined with persuasive firmness endeared him to his colleagues and subordinates and earned for him their whole-hearted loyalty. For his part, he was loyal to them and generous in looking after their welfare. On many occasions, he was quick in recognising

systemic flaws affecting individuals in matters of postings and promotions and ordering remedial measures.

Proposal on the Age of Retirement

A few months before laying down office, ACM Latif sent a proposal to the government for increasing the age of retirement of the CAS to 60 in line with the Army. He, however, made a specific recommendation that the change should become effective after his retirement!

I appreciate A.V.M. Krishnaswamy's above observations.

L to R: Idris, Bilkees, Asgar, Mariam, Asad, Air House, New Delhi, 1981

Idris mentioned a very meaningful conversation with Mrs. Indira Gandhi which I am adding in his own words as follows.

It was just a few months after Mrs. Gandhi resumed office as Prime Minister, that, after a Commander's Conference, my wife and I invited her to dinner at Air House, primarily to give her an opportunity to informally meet the Commanders, the PSOs and their wives. It was a beautiful moonlit night, and we dined with the tables out on the lawn with the roses and giant

dahlias in full bloom and the trees softly floodlit. Mrs. Gandhi clearly loved it and seemed happy and relaxed. Also, she specially seemed to enjoy the informality that we tried to maintain.

Sitting next to me at the dining table, we talked of wildlife, art, literature, family. Only once, throughout the evening, she talked to me on a Service matter. She raised the subject of flight safety, something that was clearly causing her serious concern. She said that there had been three quite serious accidents in recent months and there was growing criticism in Parliament. She asked me to make a very special effort to ensure better performance.

Prime Minister I.Gandhi presiding over Idris' last Commander's Conference as Chief of the Air Staff

Knowing that she always appreciated a frank answer, I took the liberty of saying, "Madam, I'm glad that you have raised the subject. Knowing that you like people to be frank, I really welcome this opportunity of submitting a couple of aspects that are, I believe, not always given the consideration they deserve. To begin with Madam, and if I may put it very bluntly, the time to worry is when an Air Force has NO accidents. For civil aviation, flight safety is obviously of paramount importance. But for us in the Air Force, equally obviously, combat readiness has to remain our

top priority at all times. That is our primary duty; our primary responsibility.Of course, the safety and welfare of our comrades-in-arms is of very great importance to us also. They are, after all, in an important sense, a part of our larger family. The aircrew element is, of course, exposed to much greater risk to life; and for them, it is in war AND in peace. During flying training in peace-time, not often, happily, but certainly sometimes, when there is loss of life, for all of us it is, understandably, nothing short of heartbreaking. But combat readiness must, we know, remain our central concern. And for this we have no choice but to motivate and train our aircrew to operate the aircraft to its very limits, regardless of the risk involved. And, believe me, Madam, to their credit, our aircrew never shy away."

Mrs. Gandhi listened patiently and very, very attentively, allowing me to elaborate. "Take low flying, for instance," I added in explanation, "something of critical importance in evading enemy radar detection. In times of peace, we can train the boys to fly at around 500 feet above ground level which is relatively safe. But in war, Madam, it is of little value. It has to be at a much, much lower height. Pretty risky, I have to tell you, especially at the very high speeds of modern fighter aircraft. Indeed, the pilot's own survival is at stake — and he knows it. But operational readiness, Madam, as I stressed earlier, and as we remind our commanders at all levels, must at all times, remain, our central concern."

"In the Indian environment, Madam", I continued, "a host of other factors are also involved. If you wish, I'd happily prepare a Note and brief you whenever you can spare the time". Mrs. Gandhi smiled, I felt, in sympathy, and said a brief "Thank you". She never, ever raised the subject with me again. She was truly a wonderful leader to work for.

Idris always made it a point to mingle with those present and to encourage them to speak freely and frankly. At formal meetings and discussions at Headquarters too, he would ask for the opinions of other senior commanding officers, but would say to them that he would listen to what the others had to say, but his would be the final decision, and he expected them to then follow this with complete loyalty. He also often stressed that they, in turn, must ensure that their own subordinates got full opportunity to throw up ideas. And if anyone had a particularly good idea, he should be encouraged to write it up and then ensure that it received the most careful consideration at the

appropriate level. In addition, if possible, to also give an opportunity to the individual whose idea it originally was, to participate in the discussion and to make suggestions for improvement if they had something important to contribute. Idris arranged for regular weekly meetings of senior officers to discuss points and progress. I was told recently that he received more than one such suggestion of special importance from a few officers which were promptly examined and implemented.

Quite recently, one of the officers who had, on more than one occasion, worked under Idris, mentioned to me that Idris' working culture greatly appealed to his subordinates, especially his constant endeavour to encourage them to speak freely even on major policy matters if they felt that they had any suggestions to offer. In fact, Idris often mentioned to me also that on his visits to squadrons and other units what he enjoyed very much — and often found very productive — was the time he was able to spend with the younger officers and even other ranks, if possible over a cup of tea which helped to put them more at ease. He not only enjoyed the closer human contact with his officers and men but, professionally, found it indispensable. For instance, he said to me, to get a far better feel for the morale of the Air Force at all levels.

It must have been particularly heartwarming and satisfying for Idris that, with his sensitive man-management, the feeling of fear that had crept in amongst both officers and men was gone; and, even I noticed that all — officers and men — once again felt part of a larger family. And with it, no less important, in spite of three changes of Prime Ministers, three changes of Defence Ministers and no less than four different Defence Secretaries, the very much-needed modernisation of the Air Force, particularly its strike element, was achieved.

For the final Republic Day parade of his tenure as Chief of our Air Force, a particularly spectacular fly-past had been arranged for the finale, with aircraft in the lead trailing coloured smoke in the national colours. Unfortunately, because of the heavy fog and poor visibility, the fly-past had to be cancelled. Idris mentioned the disappointment of the people to Mrs. Gandhi, and wondered if it could be added as the concluding event of the Beating Retreat event traditionally held a few days later. When she agreed, Idris arranged quick rehearsals in coordination with the Army. It was such a great success that, among other VIPs who congratulated us, Rajiv Gandhi came up and said to Idris, "Air Chief, the Air Force really stole the show today!"

Grand Finale of the Beating Retreat Ceremony, 1981

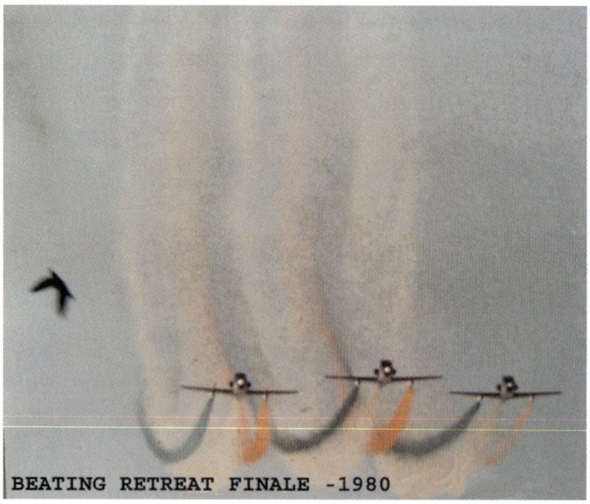

Grand Finale of the Beating Retreat Ceremony, 1981

CHAPTER 23

23 Akbar Road: Air House

Our stay at Air House was for three very active years and included my increased involvement with the Air Force Wives Welfare Association (AFFWA), as its President. I had been Honorary Secretary for several years from the inception of AFWWA and through the tenures of two Air Chiefs until we left Delhi.

Along with Idris, I travelled to the distant Air Force Stations and I was in constant touch with AFFWA members and Service personnel all over our far-flung stations. I, thus, learnt even more about their anxieties, thoughts, suggestions and positive actions in improving life in our stations. I had known about a problem that was causing great concern to Air Force parents — that of their children's education and frequent postings. There were Air Force schools established at many stations, but the schools did not have official recognition by the Education Ministry of the central government. This created a major problem for those who got transferred from a remote area to a big city. When the parents would try to put their children into a good school which had official recognition in the city, the city school would only accept children who came from a school 'recognised' by the government. If they came from just any Air Force Station school, no matter how high the standard was, they could not be admitted, as the school had not received official recognition by the government. A number of airmen's families had brought this to my notice, and their problem concerned me greatly too.

When Idris was appointed the Chief of the Air Staff and, consequently, I became the President of AFFWA, my first task was to have an AFFWA meeting and collect all the statistics on our Air Force schools and then work out the required improvements

and try to get them implemented. I then went to see the Minister for Education and requested him to get our Air Force schools examined, and if they qualified, to get them recognised by the government. He was understanding of the problem and agreed to my request and suggestion. After our schools were examined, 60 of our Air Force Station schools all over India got the official "recognition" from the central government. While we were involved with getting the Air Force schools upgraded and recognised, I was helped tremendously in the efficient coordination and implementation of this enormous task by (then) Wing Commander Krishnaswamy, who also became Idris' most able Staff Officer. He had been the Command Education Officer as well as AFFWA Officer at Maintenance Command with us and we were well aware of his dedication, knowledge and ability. Most able and dedicated in every way, for me also, as for Idris, he was a pillar of strength.

At the Air Force Day Reception at Air House, 8th October 1981

There was no question of major improvements to the house as funds were very limited in those days, but we were allowed new drapes and a carpet within a fixed budget for the drawing room. I got these in a lime green colour and created an Indian ambience rather than the more usual Western look. We sent for our own finely carved antique formal furniture which had earlier been my mother's. Our own antique Indian miniature paintings, a few beautiful old Tanjore paintings, an antique Tanka and a ten-foot-long Jamini Roy painting were placed on the walls. Our Indian *objets d'art*, with a few lovely table lamps made up the room and passage leading into it. We also had to send for my mother's former dining set which had two sideboards with antique Chinese painted panels, and a few more matching panels for the walls. Later, we were able to enclose a narrow outside verandah behind the dining room to prevent stray dogs, cats and vermin from coming inside. It was very gratifying that, when Mrs Indira Gandhi walked into our drawing room, she was most encouragingly complimentary.

When we moved into the house, the entrance passage looked very bare. It needed something special, and there was no question of buying anything in those days, so I went to the furniture depot of the PWD which provided furniture for all our government houses. There seemed to be just the usual PWD type of furniture, but suddenly while looking around I saw two long narrow side tables with elegant legs, several of which were broken. They were both piled high with other broken furniture. Thanks to my

Vice President and Mrs. Hidayatullah

Air Marshal and Mrs. Arjan Singh

Prime Minister Shri Morarji Desai, Defence Minister Shri Jagjivan Ram

earlier training as an interior designer, I could still see enough to recognise them as special period furniture pieces. When I asked for them for Air House, they replied that they were of no use as they were badly damaged. I got them to Air House anyway. They then told me that these had been at Teen Murthy House when Pandit Nehru lived there several decades earlier. That made them even more precious. I had undertaken some interior decorating assignments when we were in Delhi on an earlier posting. At the time of the Republic Day parade after the 1971 War, I had also been asked to create huge ten-foot high butterflies to be placed on top of Rail Bhavan and Vayu Bhavan, to be fluttering in the strong breeze up there, during the parade. As I was still in touch with excellent carpenters, I sent for them to repair the beautiful tables for Air House. They still grace the entrance passage there, and have now been shown in photographs in the book on Air House recently brought out by Air Chief Marshal and Mrs. Naik.

Soon thereafter, we were able to have President Dr. Sanjiva Reddy at a formal sit-down dinner on the lawns along with our senior officers and their wives. It was a very great honour. We were informed by Rashtrapati Bhavan officials that no President of India had ever been for a dinner at Air House.

President Sanjiva Reddy at a formal dinner with us at Air House

It was always a great thrill when Air House continued through the years to keep getting the First Prize each year for the Best Garden. We were delighted when it was announced that it topped the list. Happily, there it remained during our three years. After Mrs. Indira Gandhi had become the Prime Minister again, and soon after the First Prize was announced for that year, I got a call one morning from the Prime Minister's residence to say that Mrs. Gandhi would like to see the garden that morning. We were really thrilled, but it had been raining all night and the lawns were drenched, so I said to the person on the phone to please suggest to the Prime Minister to wear closed shoes so that her feet would not get muddy, and I also asked if she would prefer to come on a drier day. We were so happy when she decided to come in a little while, and along with her, Mrs. Sonia Gandhi also came, carrying in her arms a very contented and comfortable looking baby Varun, the son of Maneka and Sanjay Gandhi. When Mrs. Indira Gandhi came, she smilingly said to me that she had almost decided to cancel her visit as she received my message that she should not come! I clarified that I had suggested her wearing suitable shoes because of the wet lawns that she would be walking on, and only suggested another, more dry day if she preferred it. She smiled and said that she could not always be sure when she could manage to find the time later. We went straight on and she walked all around the garden, and I found her to be most interested and truly knowledgeable about plants. She also knew the names of the numerous varieties of flowers that were growing there.

The long and broad flower beds on the edges of the lawns were in full bloom as herbaceous borders with a variety of flowers. The low ones in front had white candytuft, verbena, phlox, sweet william, dianthus pinks, tiny yellow zinnia linearis, dainty nemesia and nasturtium, in different beds. Behind these were the slightly taller stocks, gypsophylia (also known as baby's breath), antirrhinum, larkspur, chrysanthemums, gladiolus, poppies, dahlias, celosia with the tall hollyhocks at the rear. We had colourful portulaca, which always open up in sunlight, and also carnations, gerberas as well as beautiful bunches of cineraria in different shades of blue. We had rows of salvia growing in distant beds so that the red flowers stood out. There were also long beds full of bunches of cineraria in shades of blue and in mauve, and dotted here and there were poinsettias in full bloom. There were magnificent dahlias with the largest flowers of their kind, and in beautiful colours. We also had a variety of chrysanthemums, and gladiolus in several shades.

Mrs. Gandhi mentioned that when gladiolus bulbs of different colours are planted and grown several times, the colour of the flowers changes over the years and goes back to an apricot shade. I had not known this, but we found out through the years of gardening elsewhere too that this is correct.

Below the terrace was a large flower bed with the most magnificent roses planted several years earlier, probably when Mrs. Teji Arjan Singh lived at Air House, which had later been redone and nurtured by Mrs Satya Mehra. This is where Idris began to put grain every morning for some quail that he had spotted in the bushes. At first they were too shy to come closer, but after some time of his regular spreading of grain for them, they began to come up and feed there regularly. Idris would go inside the house after putting the grain for them, and the children and Idris and I would try to watch them from inside while the quail came there to feed.

We took the Prime Minister and Mrs. Sonia Gandhi through the terrace and inside to the drawing room. I was so thrilled when Mrs. Gandhi praised the lime green colouring and some other items, especially a little side table which I had got made using an antique Andhra coconut grater on unusual carved legs as the base which were a pair of south Indian 'yadi'. We were so overjoyed by this visit by Mrs. Gandhi and Sonia, who carried Sanjay's baby so lovingly. Mrs. Gandhi was wearing a black sari, and she referred to it explaining that ever since Sanjay's death she did not feel like wearing any other colour. Neither Mrs. Gandhi nor Sonia fussed about the mud that had got caked onto their shoes in the garden, but very considerately they were careful to wipe it off on a foot mat before coming inside the house. I had often been told by the Air Headquarters Communication Squadron pilots who flew certain government VVIPs on their official visits all over the country, that whereas some VIPs — not infrequently (!) — complained about numerous things, our aircrews always said that Mrs. Gandhi, in contrast, had never complained about anything. In fact, she always, without fail, thanked the Captain and each member of the crew. Mrs. Gandhi never fussed, the crews used to tell us, and had never complained nor said a critical word even when she was really tired. On one occasion, they told me that, once, even in mid-summer, during electioneering, in spite of being unwell, on one single day, she flew by helicopter to no less than 13 places, on each flight, always polite and courteous, quietly going

through a box of files! They added that it was always not only an honour but a pleasure whenever she flew with them.

We often had functions at the house such as those when we hosted the awardees of Air Force medals and honours, as well as those who won trophies for best performances of squadrons in various fields. This gave us an opportunity to interact with the more junior Servicemen and their wives too. Our regular and frequent entertainment was for Service officers and their wives, for inter-Service interaction and government officials, particularly those connected with defence. Of a different kind of party, were two lovely cultural evenings which stand out in my memory of our stay at Air House. One was when the well-known sarod player Ashish Khan played the sarod so beautifully, with almost the magic of his famous father, Ustad Ali Akbar Khan. The other was when Sheila Dar sang ghazals in her wonderfully trained deep voice.

One emergency which came up was when there were severe floods on the outskirts of Delhi. A large number of flood victims who had lost everything had been shifted to a makeshift camp with tents on high ground. They needed urgent help as they had lost all that they had owned, and did not have the means to replace them. AFFWA members helped me with the collection of clothes. We also made packages of sets for men, women and children of varied sizes and also of household requirements. These included a duree, a bucket, cooking utensils, a kerosene stove, kerosene, even match boxes, steel glasses and plates, some rations, a sheet and a shawl and other such basic necessities. These were made into separate packages for each home. We then had them taken to the area where Group Captain Dhawan who commanded the Air Force Station at New Delhi, had arranged for a most efficient system of distribution which was a great help.

While I was busy with packing the kits and clothes, Group Captain Dhawan arranged for Service personnel to go to the affected people and give each family in the refugee camp a paper with the name of the head of the family. A copy of these names was also kept for ourselves, and so we knew how many packages were needed. When we got there with our truckloads of kits and clothes packages, we just had to call out the names and hand out a kit to the head of a family and a set of clothes to each woman, man or child. However, in a short time, people from unaffected areas came running up, hoping for handouts too, but only those who actually had lost their homes

and possessions had been listed. I was doubly grateful to Group Captain Dhawan and his team for organising it so efficiently.

I recall an incident from a few years earlier when two of our officers, who had been released after being prisoners of war in Pakistan, had come to see us. We had known them earlier as they had been with us at one of our stations, and they came with their wives. I asked one of them why, in his estimate, did the Pakistan Air Force not perform well. His reply was that although there were several reasons but, even as a prisoner he had become aware that morale was low. He said that there seemed to be a vast gap between the officers and men, and that while the Pakistani officers seemed to have a lot of privileges, there appeared to be very little interaction with, or concern for, the other ranks.

This had stayed in my mind, and the necessity for caring about all our personnel, of discussing and sharing their concerns, and of morale in general, became a matter of great importance for me. I was aware of my husband's concern and that he was working most sincerely and urgently to sort out the problem of the promotion block and other important issues. For me too, the Air Force family mattered deeply and it was important to have close bonds with them and between them. Of course, there was no question of my interfering in any way in Service matters. I was convinced however, that it was important to remind the AFFWA members about whatever benefits the Service was increasingly providing, because sometimes they were not, I found, sufficiently aware of these.

Because of the promotion block that affected so many officers who then faced early retirement, the wives of superseded officers would also write to me and come to speak to me about how their husbands had become so depressed that they would not sleep at night. The wives would see their husbands walking back and forth all night or brooding in a chair, and they asked me what they could do to help. It was a very difficult problem. I suggested that we as wives could reassure a husband in such a situation, telling him that he was the same adored and wonderful man that he had always been, irrespective of his rank, and that it was no fault of his that only a few could get promoted. Worried and disturbed wives of officers and airmen came to see me with other personal problems. As long as the matter was definitely not official, I would talk with them and try to help. I refused to hear them if the matter was about

anything official as that was not my domain. I saw to it that wives were not stopped from coming to see me in the mornings as long as there was nothing they discussed about anything connected with Service matters or postings! All that could be done in the Service was being done officially. I did not discuss Service matters at all. In any case, my husband did not discuss Service matters with me either!

However, I showed some of the letters to Idris as it was a morale problem, and then one of the letters to the Defence Minister C. Subramaniam when he sat next to me at a dinner at Air House during the period of a Commanders Conference. I knew how my husband was deeply concerned about the promotion policy and its serious fall-out. One night, I woke up to find Idris sitting up in the bed. I could make out that he was very disturbed. He told me that he was very deeply concerned with the tragic consequences of the promotional block. Some of our finest officers, including above average Squadron Commanders, he said, officers who were really the cutting edge of our Air Force, had to be denied promotions because of lack of vacancies. It was for no fault of theirs, and it was having a shattering effect on the morale capability of the Air Force. He said to me, "If the government does not accept the remedial measures we have proposed, my resignation may perhaps move them"; adding, "Just be prepared for us to move at short notice!" Happily, some relief soon came through.

Idris always made it a point to not only address all personnel, but have discussions with them too. We both told our valued audiences of the welfare measures also being worked out, and invited their suggestions. This was very important for morale in the Service. It was important for every family to feel cared for and of being considered a valued member of a vast Air Force family.

Family planning for a small family norm was something that I felt was very important. Somehow, the Class IV civilians and their families needed not only to be provided with the awareness of its benefits, but to be enabled and encouraged to practise it. For this, we worked out an effective methodology at all our stations. A very popular addition at our stations was the large number of AFFWA Vocational Training Centres as they helped the ladies and their older daughters to learn a variety of skills. At Air Force Station, New Delhi, the AOA's wife Charmaine Jena was very active in every way. We were able to establish a Polytechnic Training Centre for Women there, and to coordinate it with a very well-established and sought after Women's

Polytechnic, so that the students could get proper certificates or diplomas at the end of their training.

L to R: Mrs. Charmaine Jena, Mrs. Sikand, Bilkees, Mrs. Dilbag Singh at an AFFWA function, Palam, Delhi, 1979

The AFWWA vocational training became very popular, but the accommodation for airmen was very limited as, (we were told) due to its proximity to 7 Race Course Road — the residence of the Prime Minister — there were severe restrictions on buildings at, and in the close vicinity of, the Air Force Station, New Delhi. If anything had been permitted to be built, top priority, Idris mentioned to me once, would definitely have to be given to building additional living accommodation for the airmen which was almost tragically short. This, I soon found was so true: I had visited some on this station and I was absolutely appalled to find that in many instances, there were two families sharing a room by partitioning it in between.

Yet many preferred staying at this central location rather than to live in hired accommodation sanctioned by the government, if it was only available far away. Each Air Chief tried hard to acquire this property for the Air Force and then they could have built proper new accommodation, but the transfer of the land to the Air Force has, I believe, still not come through.

By 1981, Idris had succeeded in getting increased financial grants sanctioned by the government for the widows and dependants of those who died in air accidents. There were many who had terribly difficult family problems. Consequently, for any widow wishing to be further qualified to enable her to earn a decent living, AFFWA arranged to fund the cost of training whenever possible, and, subsequently, every effort was also made to find them work which was suited to each if they wanted to undertake such work. It was often heartbreaking for us to learn of the problems that the widows faced and to try and help in some way. Some of these still haunt me. When the government agreed to sanction better financial projects, there were other problems. In some cases, the old parents of the officer and the widow had differences, which led to litigation. Occasionally, I found, some widows were tragically being exploited by their own relatives. However, most of these very traumatised girls and their children were very courageous, even when they were totally broken within. It was so shattering for the ageing parents too. One of the hardest tasks that I faced was when I had to go and break the news to a family that their loved one was no more.

We brought out a big issue on AFFWA for which the AFFWA groups at stations contributed articles, photographs and advertisements. When I requested her, Mrs. Indira Gandhi honoured us by sending a message with a poem for it. With the funds that were raised, a hostel for girls was established at the Bal Bharati School premises. This was a great need expressed by many for those who were posted out at a crucial time when their daughters had to complete their final school year. Parents were, thus, able to leave them at a safe hostel. For boys, there already was a boarding house at the Air Force School at Subroto Park. AFFWA members put in a great deal of dedicated effort in every way. In fact, they do much more in so many ways now. AFWWA has also been fortunate in being helped by very dedicated and able Air Force officers right through. Now, as a permanent member of the IAF Benevolent Association, each year when I attend their Annual Meetings in Delhi, I feel so happy to learn about all that

Our daughter Mariam

Our son Asad

Our son Asgar

is continually being done by the Air Force non-public funds to make life easier for those who need it. For the young ones who feel that life outside the Air Force is better because it can be more lucrative, I must add here that for us life in the Air Force has remained unequalled. For the quality of comradeship and the sharing and caring, and lots of fun, are hard to beat.

During this period I was asked to be on the Delhi Social Welfare Advisory Board, which I then joined. Through this, I was able to learn about, and visit, several good welfare organisations, and also those which needed to improve a great deal. I also continued to feel that far more had to be undertaken for vast numbers of our deprived people. As Air House continued to get the Best Garden Prize every year, I felt like doing something for the Horticulture Department, so I discussed it with Idris and we gave them two rolling cups, one in the name of my mother Alys Hydari, and the other in the name of Idris' mother Leila Latif.

CHAPTER 24 Farewell to Arms

Idris' three enormously busy, but exciting, stimulating and fruitful years ended too soon. Before the end of his tenure, Idris visited all the Commands, and in the following address, he emphasised once again certain important points for all ranks of his beloved Air Force.

> *Time never stands still: we all know that for each one of us, a day must dawn when we have to say farewell. It must come for each one of us; and for me that moment in time has come today.*
>
> *But let me share some thoughts with you, thoughts about our Service, of our Air Force — about us; where we are, and the goals, the objectives, the ideals we must strive for.*
>
> *First of all, as I have often said, it is important for all of us to remember, to remind ourselves that the resources, the equipment, the facilities that our government, our nation has entrusted to us are truly of staggering proportions. They have cost the country hundreds and hundreds of crores of rupees; and they have been procured for us at a cost that the country can ill afford. It is our responsibility, yours and mine, our joint responsibility to ensure that we make the best possible use of it. Let us never forget that the strength of the Air Force — as of any organisation for that matter, depends not only on what it has, but what it does with what it has.*

Thanks to the foresight of our predecessors, their mature planning and their hard work, the Air Force today is placed on a firm foundation. Our training institutions, for instance, cover the entire spectrum of the requirements of any Air Force; our maintenance infrastructure which is in a way our backbone, is also impressive, indeed, amongst developing countries, it is unmatched; and our squadrons and other front line units, which are the teeth of our Air Force are also strong and sharp. I am sure that there can be no doubt in anyone's mind that the government is taking every possible measure to keep our weapon systems modern and up to date. But I have no doubt in my mind that our greatest asset has been, is today and will remain, the fighting spirit and the technical and specialist skills of our personnel, both officers and men, Servicemen and civilians. In skills, in experience, in ingenuity, we are second to none. If the resolve is there, the determination is there, the will is there, there is nothing that we cannot achieve. There is one other input that is necessary: hard work. Never forget that it is our capacity to work in peace and the quality of that work that will in the ultimate analyses determine the degree of success we achieve in any future conflict, should war be thrust upon us once again.

The last three years of mine as head of the Air Force family, have undoubtedly been the most stimulating and the most satisfying; and if, during this short period, we have been able to achieve anything, if we have been able to make some progress, if we have been able to add to the power and prestige of the Service, it is only because I have been able to count always on your total loyalty and the knowledge that whatever the task, each one of you, individually and collectively would do his best.

I must now say farewell to you; I say it with a heavy heart, but a great source of consolation to me is the knowledge that in my successor you have — the Air Force has — an extraordinarily fine officer, a man who not only has intimate knowledge of operations but one who is known for his humanity, his sense of justice and one who will certainly give your welfare and your well-being constant attention. And I have no doubt whatever that under his guidance, our Air Force will continue to grow from strength

Idris' farewell salute at the formal parade in his honour

Idris and Bilkees at their farewell at Air Force Station, Agra

Farewell at Jodhpur, South-Western Command. Being towed by a MiG-23, July 1981

Final drive through Air Force Station, New Delhi, on eve of retirement, 30th August 1981.

Bilkees: Farewell visit to Jodhpur, 1981

A warm welcome to Air Chief Marshal Dilbag Singh, his successor, as the new Air Chief

to strength; and by its actions and its accomplishments, its devotion and its dedication, by its prowess and its performance, add lustre to its name.

Idris and I vacated Air House the morning after his retirement. We were fortunate that the government had allocated 5 Teen Murthy Marg for us to live in for six months, as the Principal Secretary to the PM told him that the PM wanted him, immediately thereafter, to serve as a member on the Public Services Selection Board. On Idris' first day of retirement, what brought the fact home to him, he told me (and has been mentioned earlier) was when he went to dress in the morning. From the time that we got married, right through the years, I had always, each morning, laid out his uniform for him to dress. But that morning, when he went to dress, it was not there! I have to confess that I never recognised the different rank badges as I only looked at the current ones which I fixed onto his uniform! That first day of his retirement, he said that he felt it was so unreal and strange not getting into his beloved uniform. He told me later, that it was then that, with a shock, he suddenly realised that his days in the Air Force that he loved so much, were over.

We had decided to vacate the house on the day that he retired so we were all set to leave Air House after breakfast. We had a last look around at the house and garden. Idris then got into the driver's seat of our private car and I got in beside him. We drove out of Air House with a farewell look at the last Air Force home and garden that we had been so privileged to occupy and care for. We had lived in 25 different homes at different stations in our Air Force life after we got married. Idris had spent 40 years in his beloved Air Force, of which nine were as a bachelor! Our furniture and other belongings had already been sent out the day before. I wanted to leave the house clean and neat for the Dilbag Singh family who were to move in. We wished our successors and the Air House a happy togetherness.

It was quite funny that as Idris drove out of Air House with the car loaded with odds and ends and me seated next to him, and about half a kilometre on, a tyre burst and the car went, 'thud thud thud'! Idris, knowing Dilbag, his successor's personality, had no hesitation in calling him on the phone, saying laughingly, "Dilbag, I never thought I would have to lean on you so soon, but my private car has had a puncture on our way to our newly allotted house. Would you please send an Air Force car to pick us up? What a way to start retirement, isn't it"? Dilbag, being the kind of man he

was, a car arrived within minutes, and we soon reached our next home. I had already arranged our furniture with flowers in vases too, so it was welcoming and comfortable.

Idris felt satisfied as we drove out of Air House on his retirement, that some of his major objectives had been achieved: morale was good, a unique modernisation of the weapon systems had been substantially implemented and he had been able to fly even the latest acquisitions, including the MiG-25, shortly before.

After his retirement from the Air Force, whether he was Governor of Maharashtra or Ambassador to the fabulous country that France is, or even after full retirement in more recent years, Idris has continued to be deeply concerned in the improvements and effectiveness of his Air Force. When he was Governor of Maharashtra, for instance, and was told by his successor, (the late) Air Chief Marshal Dilbag Singh that certain benefits which had been approved by the Prime Minister before he took office, were still awaiting financial sanction of the Finance Ministry, Idris spontaneously offered to intervene and pursued it on a personal basis with the Finance Minister, the (late) Shri Venkatraman who got final orders issued within the month.

What is particularly heartwarming is when we come across officers and non-commissioned ranks who see Idris somewhere and come up to greet him. They invariably mention that they can never forget that he cared for personnel so much. Even now, Idris sometimes tells me, "I hate adding 'Retd.' to my name; but I have to say that I am so very, very proud of our wonderful 'Air Warriors' as they are now called; of every rank, commissioned or non-commissioned, who, whenever we are together, always make me feel that I am as close to them as I was more than a quarter of a century ago."

CHAPTER 25 Governor of Maharashtra

On 1st September, on Idris' retirement, the Prime Minister issued a press notice (as reported by PTI) and sent Idris the letter reproduced in the early part of this book, highly appreciating the manner in which Air Chief Marshal I.H. Latif headed the Air Force, raising the morale of its officers and men. I was deeply touched that she also praised my role in it.

Soon after Idris retired, Idris was made a member of the Public Services Selection Board and also of the Indian Airlines Board. Two new assignments were also suggested to him. One was as Chairman of HAL (Hindustan Aeronautics Limited), which he said that he would be happy to accept. But, he added, that HAL now had innumerable branches spread across the country, and he had learnt from the present Chairman that he had found to his dismay that he had, perforce, to repeatedly visit Delhi as almost all important policy decisions were almost invariably by prolonged discussions with ministries of the the central government. Keeping this in mind, Idris added that he would want to shift his office with supporting staff from Bangalore to Delhi. The Principal Secretary to the PM, Dr. P.C. Alexander, whom Idris greatly respected and liked and to whom he had spoken, said that this should be acceptable. But within days, Dr. Alexander called again to say that the Prime Minister now had other plans for him. Meanwhile, as an interim measure, he was being appointed as a member of the Public Services Selection Board and being allotted a house on Tughlak Road commensurate with his appointment. We could, for the time being, continue to live in Delhi.

Perhaps four months later, Giani Zail Singh, the Home Minister, called Idris on the phone to say that the PM (Mrs. Gandhi) had said to inform him that he was to be the Governor of Maharashtra. Idris said that it would be a very great honour, but for this, as the existing Governor, Air Chief Marshal Mehra under whom Idris had served and whom Idris very highly respected, would have to be transferred, he requested that the change be deferred till the existing Governor's term ended. Idris really felt very, very deeply about this as he had also served for years under Air Chief Marshal Mehra and both he and his charming and most affectionate wife had also been very caring elders. However, Idris was told that Governor Air Chief Marshal Mehra was only being moved to Rajasthan as the Governor of that state. Concluding this, the Home Minister told Idris that he would have to move almost immediately to Bombay as the State Assembly session was due to start within days, and he would be required to address the opening.

Formally taking over as Governor of Maharashtra

The following account is from a note that Idris had written soon after our return to Hyderabad, and had fortunately preserved.

The Prime Minister, Mrs. Gandhi had decided that I should take over as Governor of Maharashtra as soon as possible, and definitely in time to address

the forthcoming Assembly session of the State Assembly on the 6th of March. It was totally unexpected; and whilst I felt greatly honoured by the trust being reposed in me — particularly as the political atmosphere in Maharashtra had been rather destabilised by the exit of its Chief Minister, Shri Antulay, I would be less than frank if I do not confess to a feeling of some uncertainty if I, with my 40 years in the Air Force, would be able to do justice to a primarily political responsibility. But, of course, I agreed, especially as the direction came from Mrs. Gandhi herself, a person for whom I had developed great respect and regard, particularly during the period when she was Prime Minister and I had been the Chief of the Air Force. After conveying my acceptance to the Home Minister, I broke the news to my wife that I would be leaving in 2-3 days, but adding that she could follow after a fortnight as she would have to shut house and pack all our personal belongings.

Guard of Honour at the airport on arrival in Mumbai as Governor of Maharashtra, March 1982

On my arrival 2-3 days later in Mumbai, I was given an exceptionally smart Guard of Honour by a contingent of the State Police. However, since I had retired from the Air Force and was in civilian clothes, I decided that it would be improper to respond with a military salute, and the correct response would

be the nationally recognised and exceedingly dignified Namaste. This, in my considered view, was — and remains — the only proper form of salutation for civilians, especially on such occasions. I have to say that in later months, when some close friends would remark at my form of returning the salute, I would respond by saying that I fully recognised that for such and similar occasions, Pandit Nehru had himself given the lead to civilians by an attempted military salute. However, I had always felt that the senior Defence Services and civilian officers around Panditji might have had the sense and courage to advise him that a salute was a purely military form of greeting and in NO country in the world (including Britain, whose traditions in such matters we had adopted) did civilians, even at military, or other such occasions, ever salute. This was a clear indication to me, at least, that for our country, a Namaste was, and is, the most appropriate and dignified form of salutation by those not in uniform. I was also quite convinced, I would add, that had this been brought to Pandit Nehru's attention — as it should have been — Prime Minister Nehru himself would, without hesitation, have accepted their submission. Perhaps, I should add, that I have consistently represented this to many senior dignitaries and continue to hope to this day that India too will, one day, adopt, international practice in this respect.

Idris and Bilkees on their first day at Raj Bhavan, Mumbai, 1982

Almost immediately on my arrival in Maharashtra, I was faced with a difficult and politically sensitive challenge. Shri Antulay was the Chief Minister during my predecessor, Air Chief Marshal Mehra's tenure as Governor of the state. Shri Antulay was reputed to be strong-willed and, at times, with a somewhat abrasive way at work. To make it worse, many aspersions had been cast on the ethics of some of his actions as Chief Minister. Not unexpectedly, a prominent figure in the Opposition, Shri Ram Naik filed a case for his prosecution on criminal charges, and the Centre soon replaced the Chief Minister. On arrival, I was briefed that before Shri Antulay could be tried, it was the Governor's direct responsibility to personally and entirely on his own, determine whether there was a prima facie case. I had, of course never even seen the Criminal Procedure Code, leave alone understand it, which made the challenge a truly daunting one! The examination of the charges, I knew, would be a laborious one as well, as no less than 13 charges had been framed. This meant, I soon found, that the law would not permit my seeking advice. I would personally have to, most carefully, scrutinise every enclosure and every noting in, perhaps, each of around a hundred files.

For those three-odd months I really burnt the midnight oil. By mid- July, I had been through all the related government files and documents of the 13 charges that had been levelled against Shri Antulay. Based on what I saw, I was convinced that in five cases there was prima facie evidence of guilt. Several Ministers who came from Delhi suggested exoneration. I understood their interest, but requested them to accept my compulsions as the law required me to take the decision entirely on my own and without advice! But knowing the compulsions of these Ministers also, I would always add that if there were certain compelling reasons why they felt that the decisions must go a certain way, I would willingly resign from the Governor's post.

Thereafter, I spent two days at the Pune Raj Bhavan to finalise and dispatch my assessment and findings. As required, apart from the government and Shri Ram Naik (of the BJP), I sent a copy to Shri Antulay also. Soon the latter requested a meeting which, of course, was arranged on my return to Mumbai, particularly as I could imagine his distress. Sadly, when he met

me, he was very, very bitter. He was clearly convinced that I was following the Prime Minister's direction and mentioned that he would be meeting the press in a few days to share the 'truth' with the people .Of course, the Prime Minister had never spoken — or ever even sent any message to me on the subject. In fact, when I was in Pune finalising my report, she was in the US.

At this meeting with Mr Antulay, I told him that I had been informed that he (Mr. Antulay) intended to call a press conference and that there would be accusations against the PM. I said that my intention in inviting him was only to convince Mr. Antulay that Mrs. Gandhi had been very correct throughout, and had absolutely no role to play in the findings. I added that, for his sake I was, therefore, anxious that the press conference should not be based on any misconception in this regard. I also said that I wanted to protect Antulay from making accusations based on incorrect assumptions. Mr. Antulay was clearly deeply disturbed, and he said, "Governor Saheb, when I go to court you will be the first witness that Barrister Antulay will question." However, thereafter, the press conference was called off by him. Coincidentally, Mrs. Gandhi made an unexpected visit to Bombay soon thereafter when Amitabh Bachhan was seriously unwell. When I met her, I asked her if she would like me to give her a brief report on the findings. She said, "No, no. I have seen your report to the Home Ministry. I am sure you have done what you felt was right."

A long time ago, when I was a boy, I had requested Sarojini Naidu who was a close and dear friend of the family to sign my autograph album. In it, she had written: "Of all the qualities of the human mind, the one I value most is courage: not only the courage to face danger, but the courage to obey the dictates of one's conscience in doing what is right."

Several years later, Mr. Antulay went on to become a Union Minister and a good one too. Bilkees was a member of the Governing Board of the National Red Cross and was at Governing Board Meetings at which he was presiding as the Health Minister. She found him to be cordial and most efficient at that period. He also invited us to his daughter's wedding, which was gracious of him, and we were happy to attend.

When I assumed office, Babasaheb Bhonsle was then the Chief Minister (Congress). I suggested to him how he could build up a greater base in the party by having MLAs who strongly backed him. Babasaheb said, "How can I when they only come to me when they want a liquor licence and allotment of a wine shop"; and he showed me a list of those who had actually asked for these! I said that he should inform Mrs. Gandhi about this. I added that he should also maintain a chart of the MLAs and their involvements and activities, of what they come for which has been of benefit solely to themselves, rather than for constructive state development or nation-building work, and then show this to the party leader.

Happily, as a person, I found Chief Minister Babasaheb Bhonsle most courteous, warm and receptive; and, very quickly, I was able to establish a comfortable and easy working relationship with him. And to me, he also appeared both straightforward and a loyal member of the ruling Congress Party. Though he was conscious that his appointment quickly following Shri Antulay's exit was, in a way, a stop-gap measure, he clearly tried his best to meet the challenges facing him. One was that he didn't seem to have a strong lobby even within the party, partly, in all probability, because others were also aware of the seemingly temporary nature of his appointment. To compound his problems, that year, the state was facing a second year of very severe drought and, as a consequence, financial and related pressures were obviously very serious. During one of our early meetings, I mentioned to him that the annual large and rather lavish Independence Day receptions hosted at Raj Bhavan (traditionally held in Pune — the one on Republic Day being at Mumbai) would for that year at least, not be appropriate, and suggested its cancellation. I suggested also that publicising the reason would be well-received by the public and do him credit. He thanked me and agreed immediately. Happily, public reaction was all in favour, and the media, in fact, greatly applauded the action and his sensitivity.

I have to say that I found this assignment not just challenging but, emotionally, most satisfying. In a sense, I also felt that it was a great education because it brought me close to a wide cross-section of our wonderful people, including those in our rural areas, people with whom,

whilst in the Air Force, I had little or no intimate interaction. It was easy for me to develop a warm understanding with a lot of the political figures, with the Chief Ministers as well as those of the opposition. Particularly challenging — but also very satisfying for me — were my responsibilities as the Chancellor of the 11 universities of the state. One particular challenge that I faced, almost from the very day that I arrived, was when I discovered that Bombay University, which enjoyed the reputation of being amongst the very best in the country, while holding their exams on schedule in May/ June but shockingly, were declaring the results several months later.

Clearly, this was unacceptable. At this time, Bombay University was to celebrate its centenary. I wanted to call off the celebrations or the big party that they were planning. I, therefore, decided to inform the Vice Chancellor that I, in my capacity as Chancellor, would be chairing the meeting. This was, I was told later, quite a shock to all those who had assembled, but they all knew that, constitutionally, I had every right to do so. I spoke at some length, but my main theme was simple: "What are we celebrating when the results of exams for which we are directly responsible are held as normal, in May, but the results, which, we well know, are of critical importance to the students, are, year after year, being declared months later towards the end of the year?" However, I also immediately followed by adding that, to start with, I, with them, all together, should do our utmost to get the results on time and hold the celebrations as soon as that is achieved. Then it would be for all, staff and students alike, a true celebration. With a sense of relief and satisfaction, I noticed that almost without exception all smiled, and most were even nodding their heads. Clearly, there was no resentment.

On my return to Raj Bhavan, I took steps to find solutions to set matters right. I first asked the Vice Chancellor to come and meet me urgently and, when he came, told him that the inability of the university to declare the examination results in time must, clearly, be having disastrous effects on the careers of the students at a critical stage of their lives. This, I went on to say, was totally unacceptable to me and must be set right on very high priority. The Vice Chancellor responded, rather apologetically, saying that it was an impossible task as the number of students had risen to

unmanageable levels. I was shocked and deeply disappointed, but with some sadness in my heart, I very firmly said that if he felt that it was an impossible task for him, he was leaving me with no option but to ask him for his resignation. This, understandably, deeply distressed him and he almost broke down in tears. Seeing this, I quickly added that I had an alternative: he and I, I said, together with a specially selected IAS officer (whom I would request the state government to make available to me on priority), would form a team to study the problems in depth, again on very high priority, and try and evolve appropriate remedial measures. I felt confident, I said, that we would be able to identify steps that would substantially improve matters. The VC was greatly relieved and, thanking me, promptly agreed.

Happily, the state government, also immediately, placed a relatively young but outstanding IAS officer, D.T. Joseph, at our disposal. At the initial meetings, both I and the Vice Chancellor made some suggestions, but largely, very largely due to D.T. Joseph's truly outstanding work, including his assessment that by largely adopting the current system at the Pune University, the entire system would become effective. I instructed the Vice Chancellor to implement it. Steps were additionally taken for corrections of exam papers by taking on extra, paid personnel for this. They were to come to the university from 9–5 to correct these papers. I said that for this work, they could take on those who had retired from the university within the previous two years. The numbers increased immediately of those correcting the examination papers, and results

At Kolhapur, with a Temple elephant that was particularly welcoming

thereafter were out as scheduled for the next exams in June 1983, on time, almost entirely, I have to say, due to the excellent work of Mr. D.T. Joseph.

My appointment as Governor gave me a most welcome opportunity to tour and familiarise myself with the problems and promises of the rural areas and our villages, where, as our Mahatma Gandhi, memorably said "the heart of India beats". Regularly, therefore, on my instruction, visits to the rural areas were incuded in my monthly schedule, and whenever possible, my wife was always delighted to join me. For my quite frequent visits to the rural areas, I however, usually asked the Commissioner/Collector NOT to arrange our lunches (my wife's and mine) not at Government Circuit/ Guest houses but, if at all possible, at a village, with the Sarpanch or senior elder of the area and his family. Sitting and chatting with our simple hardworking, kind, generous — and often very mature and knowledgeable — people in our villages was not just a novel experience but always a most heartwarming and memorable event; and often, for me, an education in more ways than one.

One memorable afternoon, I still vividly recall, was when my wife and I were having a very informal lunch/meeting, sitting on mats spread under a tree with the Sarpanch of the village and his family. The simple lunch over, the Sarpanch asked me if we would consent to be photographed with his family. Touched and delighted, we, of course, immediately agreed. Soon after the first photograph, the Sarpanch hesitatingly asked if I would consent to another. Though a little intrigued, I, of course, readily agreed. Lo and behold, two superb animals, a cow and a bull were brought to join us! For my wife and for me, it became one of the most touching and memorable moments of our lives. It also brought home to us in a wonderful and truly unforgettable way the close and loving bonds that bind our people to these wonderful animals; these wonderful gifts of God to us all. Later, back at Raj Bhavan, I thought to myself that had my lunch been arranged by the Commissioner at a Circuit House, how great my loss would have been!

Another important lesson brought home to me by my not infrequent interaction with our farmers in the rural heartland of our country, was that the fruits of

research and development in the key area of agriculture, remained woefully inadequate. My visits to our five agricultural universities, consequently acquired a special purpose and urgency. I soon also learnt that, fortunately, all our five agricultural universities had well established research and development wings. Thenceforth, I was determined to do my very best to promote greater interaction between the universities and the farmers. I soon learnt that what were known as 'Extension Programmes' were, in fact, being actively pursued. But clearly, to me at least, both the pace and contents needed urgent attention to enhance their effectiveness. To encourage this, I initiated regular formal meetings with the Vice Chancellors at Raj Bhavan, when promoting greater interaction between them and their officers, particularly those of the research and development wings, and with our farming community, became a key objective.

Another initiative that I felt compelled to take during my tenure in the field of education (though not higher education at the university level) was concerning the urgent need for an in-depth review of the syllabi for schools at all levels, both its quantum and content, keeping in mind particularly the vast differences both in the needs and capabilities of the students of rural and urban areas.

This was convincingly brought home to me during my tours of the rural areas. During such visits, I had directed the officials concerned that my informal visits could certainly be planned in advance. However, specific schools, Public Distribution System (PDS) units, government health care facilities, etc. should not be selected and alerted in advance as, when this was done, elaborate and very special preparations are made. As a result, I would never get a realistic picture of the normal every day routine working of the institutions.

Consequently, during my visits to rural areas, whilst driving through such habitations ,including small towns, I would, without notice, ask my official escort to drive up and stop at the nearest school of that area. I would then occasionally slip in by the rear entrance and, from the back seat, observe the teaching that was being conducted. Sadly, on far too many occasions, I observed that there was a near total disconnect between several students and the teacher. At the

end, I would, very privately discuss the problems with the teacher. What stood out, most often, was that for rural children, whose support systems were almost non-existent, the syllabi were: (a) by far, too much to pack satisfactorily into the time available during the year; (b) for the environment in which they spent their lives, the contents often had little relevance to their daily lives or future needs, and often, as a consequence, were not easy to comprehend. For instance, when teaching about animals, the animal in the textbook was a zebra! Surely a cow or a dog or a cat would have made more sense?

Furthermore, sadly, little importance was given at many schools to organised games and sport and other extra-curricular activities. Finally, largely, I felt, because of my unannounced visits, I detected frequent teacher absenteeism. One tragic cumulative effect of all these factors was, I was convinced, an alarmingly high rate of drop-outs. Happily, the vast majority of teachers was very conscious of these deficiencies/ limitations, and would represent to me that drastic change, in both size and content of the syllabii were urgently called for. There was, also, no doubt in my mind (and several teachers would agree) that, for our country, and certainly at this point of time, and this stage of national development, the vast differences in the needs of the rural and urban schools would have to be squarely and urgently faced.

Any major reorientation of the syllabi to satisfy the needs of both the rural and urban students was, clearly, as I have said earlier, unrealistic. Nevertheless, I frequently discussed these problems with the government, at both ministerial and official levels. Concurrently, thereafter, I also started regular visits to the concerned institution in Pune where, I was happy to find that almost all were fully conscious of the need to bring about important changes. During our interaction, several changes of significance were not only noted with enthusiasm, but, I was greatly encouraged to find that, in fact, some changes were being implemented even before I left the state in early 1985.

I think it was in May 1984 that Idris was requested to address the Tata Institute of Social Sciences. He was really delighted and, as he confessed, he was flattered too, and it

gave him an opportunity to publicly express some thoughts that he gave importance to.

Address at the Convocation of the Tata Institute of Social Sciences Bombay

Mr. Bhabha, Dr. Desai, Members of the Board of Governors, Members of the Faculty, Ladies and Gentlemen: I cannot, I fear, even by a good stretch of imagination, claim to be a social scientist and, therefore, I can only presume that in inviting me to deliver the Convocation Address at this very prestigious institute — I can only presume — that you have taken a carefully calculated risk! However, whatever be the reason, for me at least, it is a welcome opportunity to speak to what I consider a very distinguished cross-section of our society.

Today, this institute occupies not only a unique position amongst our institutions of higher education, but it is also one that has established for itself a most enviable reputation in the field of social sciences. The great demand for your graduates in industry, in commerce and in social service and other allied fields is proof, if proof is needed, of the standards, the calibre and the quality of the training that you provide. Like other institutions established by the House of Tatas, this also bears the hallmark of excellence, and small wonder that even 20 years ago, the Government of India gave the recognition to this by declaring it a deemed University.

From your prospectus, I find that the primary thrust of your work is the teaching of social sciences with a view to providing professional personnel in social service work and in personnel administration and allied professional fields. It is I think in keeping with the highly professionalised needs of our times that social work is also now labelled as a profession because in the part, indeed from time immemorial, it has been considered little more than an avocation of the kind-hearted and charitably inclined. In fact, I would say, social work today has not only graduated to the status of a profession but it has become a profession with a difference. Whereas in most professions, the primary motivation is monetary gain and the status and power that go with that position, in social work, the important motivation, indeed the key ingredient, I may say, is service. For this reason, workers in this field, in

my view, belong to a specially distinguished group of individuals, a group with a mission in life — the mission of helping fellow human beings achieve a better quality of life, a life of purpose and of dignity in an atmosphere of social equality.

Here in Maharashtra, we have so many institutions which are engaged in social work; so many who are always willing to lend a helping hand. What should be the important planks on which the work of such societies and institutions rest? It certainly should not be just charity and it most certainly should not be pity. Mind you, these are by no means unworthy sentiments, but I think all of us know from personal experience that these sentiments, however good they may be, are to most recipients, rather humiliating.

I am reminded in this context of what a Jewish sage once said. He had outlined eight grades of charity. He said the first is to give reluctantly, the second is to give cheerfully but not sufficiently, the third is to give cheerfully and sufficiently but only after being asked, and so on. The eighth and the highest grade, he said, is to provide a fellow human being with the ability to earn his own livelihood and thereby forestall the need for charity. The best service that we can render to those in need is to provide them with the training, the care, the guidance and the opportunity for work so that they become self-reliant. In every society, all over the world, economic productivity is amongst the most important criteria of achievement; for the needy, it goes further: economic productivity provides self-respect and a sense of pride. It is only then that they can lead a life of dignity, of self-respect and of optimism, each so vital for life to be worthwhile.

What is the true measure of the strength and prosperity of a nation? Its affluence or its revenues is, of course, one measure, its military strength another, some are also inclined to take into account the splendour of a nation's public buildings and institutions. But I think that all of us are aware that the best measure of the greatness of a nation is the quality of each generation that comes of age. In this process of growth — or decay! — education, we all know, plays a central role; without it no nation, no people, can prosper or even survive. Happily today, in our country, people are awake and conscious of

the role that is their right to play and with it also the realisation that poverty, deprivation, degradation — none of these is either inevitable or unchangeable. In our striving for the betterment of the lot of the common man, we have to constantly remind ourselves that it is mainly through education that we can promote equality and social justice. It is mainly through education that we can reduce the social tensions and cultural differences. Education is too vast and complex a subject to be adequately dealt with on such an occasion, but I cannot help but at least highlight one or two aspects on which I feel rather strongly.

We often tend to forget, for instance — happily, this institute does not — that as Mahatma Gandhi once said: "The heart of India beats in its villages". In determining the content of our education and related activities, and this includes higher education, we often tend to overlook the fact that 70 percent of our population is still a rural population — not only their outlook but their needs are also different. It is also necessary to keep in mind that it is only on their prosperity that we can build the prosperity of our country. And yet our education as a system still remains, in my view, highly urban-oriented. Education must have greater relevance to the needs of our rural population. Furthermore, schools and institutions of higher learning must also provide remedial programmes for students coming from backward communities and backward areas so that they also have comparable opportunities of improving their lot.

Whilst talking of education, I think many now realise that our system lays far too much stress on theoretical and academic excellence. These are, of course, important, but is not character also an important part of human resource? Can education afford to leave development of human character and personality as a by-product of intellectual development? We, of course, recognise the key role of character, but thereafter there is seldom any follow through. Activities which we also know have tremendous character-building potential are given, at best, a secondary role in our curricula. For instance, activities such as sport, social service, NCC — these I strongly feel must not only be made an integral part but a most important part of our curricula, for it is activities like these that directly and most effectively develop qualities of loyalty and leadership, qualities of grit and determination, of discipline and team spirit, of harmony and cooperation;

and, woven into all these, the genuine concern for the welfare of fellow human beings. No level, nor any field of education must ignore the key role of activities such as these.

Development and transmission of knowledge is no doubt one of the main functions of our institutions of higher learning. But, at the same time, it is necessary to ensure that knowledge that is created is relevant to our life situations. Apart from the rural bias which I have already highlighted, there is another aspect that is particularly relevant as far as social sciences and social service is concerned. We have to recognise that today no less than 40% of our population continues to live below the poverty line, with little to eat and often nowhere to live. To make it worse, we have a disabled population of no less than 60 million. When we talk of development, when we talk of social change, when we talk of improving the quality of life, it is these people — the poor, the sick, the handicapped, the illiterate, the underprivileged — that we have to keep uppermost in our minds. These are our weakest links and it is these elements that we must seek to serve and to strengthen if the edifice of our nation is to become truly strong and stable.

It follows then that our studies and our research is based largely on these people, their problems, their trials, their tribulations, their hopes and their aspirations. The results of our studies and our research must, of course, then become available to our policy-makers and planners also so that national policies and national goals are action-oriented programmes and truly reflect the needs of our country and our people. In the bargain, the students, the younger generation also gets a better insight into the needs of those who are less fortunate.

When we talk of realistic and action-oriented programmes, we must, of course, be very clear in our minds about the basic cause or causes of the problems that we face and the dimensions of these problems. It is only then that we can correctly determine the strategy that we should adopt. This is true of every field of constructive activity; it is true of social work also. What then is the basic cause of the problem? To me, it appears that the basic cause, certainly in our country, is the tremendous growth of our population, coupled with the paucity of

our resources. It is the sheer pressure of numbers, the unending growth of our population. Since Independence, our already enormous population of over 300 million, already steeped in poverty, has within a few decades, jumped to the almost astronomical figure of around 700 million. And what is of even greater concern, I am tempted to say, even more frightening, is that in spite of a major and continuing effort on the part of the government, in spite of substantial inputs of resources in practically every field of welfare activity, the problem of poverty is no better, perhaps worse; inequalities have widened. As I have already said earlier, nearly 40% of our people still continue to live below the poverty line. I have no doubt in my mind that this is the basic cause of the suffering that we see around us. If proof is needed of what havoc pressure of population can create in the lives of individuals and society, one has only to take a short trip to Dharavi. When I went there the other day and spent an hour or so, it was not so much the spectacle of poverty that shattered me, but the filth and squalor of the places that these poor people have to call their homes, their homes because the vast majority of them at least, have nowhere else to go. And what is the result of such pressure of population and the result of the intolerable conditions in which the people have to live? The result is — and I was told about this by one of the officials who accompanied me and who works in that area every day — that one out of every three is either a T.B. patient or a leper. That, ladies and gentlemen, in a nutshell, is where the pressure of population is leading our country. The conclusion is inescapable: unless we tackle population, we cannot tackle poverty and unless we tackle poverty, we cannot deal with the sickness and sorrow that is the primary concern not only of the government but also of all of us, including institutions such as yours and the social scientists and social workers that you train.

To those of you who have received your degrees and diplomas today, I offer you my warm congratulations. But may I add that you must realise that having it within your grasp, but always with the spirit of service as your guiding star, think not only in terms of what one can gain in life but also in terms of what one can give to others. On this unique and memorable occasion, let this be your promise to the people.

CHAPTER 26 At Raj Bhavan: Administration

Governor's Escort

The following is also from Idris' note which, as I have mentioned earlier, he wrote (and preserved!) on his final return to Hyderabad.

> *Within a week of my arrival in Maharashtra, I began to feel that, for my escort, the number of cars plus the police pilot jeep was, embarrassingly and unnecessarily large — and also rather ostentatious, at least for my temperament. After discussing this with the authorities and departments concerned, government accepted my request that just one police motorcyclist should precede my car, primarily to ensure that I was always dead on time for every appointment/engagement. I stressed that the person selected would have to keep in mind that I was exceptionally particular about punctuality and a plus or minus two minutes bracket only would be acceptable to me both for social engagements and for public functions. I was also quite emphatic that normal traffic must never be stopped. And my escort should strictly observe normal traffic rules, using the siren only in an emergency or after obtaining my permission. Certainly, this was very demanding, I knew, but, over the years, as an ordinary citizen, having often suffered, I was, at the very least, determined to try and ensure that my movements did not cause unnecessary inconvenience to the public. When the selected police officer first reported for duty, I personally briefed him also on my rules — and my temperament! To his credit, he responded*

with a smile and a smart salute. To his even greater credit, throughout my three years there, was that there was not a single occasion when I was too early or late. I was so impressed by his performance that, apart from giving him a personal letter of commendation, I also wrote to the Chief Minister to reward him suitably for his truly outstanding performance throughout my three years.

Policing and Helicopter Operations from Raj Bhavan

Soon after my arrival at Raj Bhavan in March '82, after my swearing in, as expected, there were many callers, including officials. Amongst them was Mr. Rebeiro, the Chief of Police in Mumbai. I was particularly taken by his forthright but always pleasant and respectful manners and, since the police is always a vital arm of the administration, I encouraged him to lean on me in times of need. At one meeting during the early months, he mentioned to me that, although his force was doing its best, he could serve the state far better if the strength was increased, commensurate with the tremendous increase in the population of the city, coupled with the vastly increased requirements of VVIPs and VIP security. Illustrating the point, he said that, for instance, for every visit of the President, Vice President or Prime Minister, for both arrival and departures to and from the airport and Raj Bhavan, no less than 700 policemen had to be deployed. The very frequent arrivals and departures of other VVIPs and VIPs, thus, naturally put a very great strain on his already very limited resources. As a consequence, for prolonged periods, the policing of several areas became totally inadequate.

I knew that he spoke with great sincerity and there was much truth in what he said. I, therefore, undertook to speak to the CM and the Home Minister, if necessary; but adding that I would also try and think myself of measures that could, perhaps, prove helpful. Whilst speaking, I had in mind the use of helicopters — though I deliberately didn't mention it to Mr. Rebeiro until I was satisfied that I had in mind a helipad at Raj Bhavan. Having done many years of flying in the Air Force, though mine was largely on fighters, bombers and transport aircraft and not on helicopters, I was very clear of

the basic safety requirements of operating these as well. The very next day, therefore, I carefully surveyed the very extensive grounds at Raj Bhavan, particularly the extensive lawns on the seaside of the main building and the even larger open grounds across the main internal road fairly close to the Raj Bhavan (upper) entrance/exit. After careful consideration, I was satisfied that a helipad for light helicopters like those already in use for VVIPs by the Air Force, was definitely a possibility, particularly at the latter location. However, keeping in mind the need to enlist expert opinion, I consulted the senior-most helicopter pilot of the state's small fleet for air communication. Incidentally, I had done a couple of flights with him earlier and had been impressed by his maturity and flying skills, and felt very comfortable consulting him. He fully agreed with me that very safe take-off and landings were definitely feasible and, keeping in mind the generally prevailing wind speed and direction and the flat vacant area near the entrance gate, a very safe landing and take-off approach lane for the helicopter could be ensured.

Since the safety of VVIPs and VIPs and important officials would be involved, as an added precaution, I then personally briefed the Chief of the Air Staff, the late Air Chief Marshal Dilbag Singh, an officer who, incidentally, had succeeded me and someone I knew exceptionally well. I requested him to assist me by sending to Raj Bhavan a very carefully selected light helicopter pilot for a couple of days to independently assess the feasibility and safety of our tentative plan, and advise us. He sent me two! I personally explained to both our requirement and our initial assessment, covering also, at some length, all the factors we had taken into account. Their questions to me, I remember so well, gave me added confidence in their competence. They fully concurred with our initial assessment. A couple of days later, I asked Mr. Rebeiro to see me and it was a joy to see his delight at our plan. We then informed the state government and, with their concurrence, the helipad, together with an adjoining crew room was soon constructed. Thereafter, after a few trial flights, in two of which I also went, to the delight of all — including the then Prime Minister Mrs. Gandhi and the Chief Minister Shri Vasantdada

Patil — helicopter communication flights for VVIPs and others to and from the airport, directly to and from Raj Bhavan, were formally authorised, and gradually came into regular use. On recent visits, I noticed with great satisfaction, it is in quite intensive use!

Renovation/Reconstruction of Raj Bhavan

Before I left Delhi to take up my appointment, the Prime Minister, Mrs. Gandhi, had mentioned to me that, in recent months, my predecessor had often stressed that some of the Raj Bhavan buildings in Mumbai, including — in fact, particularly — the main one, had, over the years, deteriorated to an extent that it was no longer safe and should be replaced on priority by new construction. The total cost of the work tentatively approved was Rs. 1.25 crore. The Prime Minister was clearly unhappy that a pre-Independence building, and, therefore, of some historical importance, would be lost. She, therefore, asked me to make a fresh and final assessment before the government agreed.

Within weeks of my arrival I, therefore, made an initial personal assessment and noticed that there were definitely quite a few large cracks in the floorings of a few rooms and leakages in some others. The main building had also, from time to time, clearly undergone additions of rooms and alterations under succeeding Governors, and, from the outside at least, had, in my view, become quite unattractive. The main original structure, however, looked pretty solid. Consequently, some good external remodelling plus good internal repairs could be considered. However, in view of the strong recommendation that had already been made by the state engineers that the building was no longer safe, I felt that they may be reluctant to reverse their findings. I, therefore, decided to request the local Specialist Army Engineering Unit to carry out an independent assessment. After several days of quite intensive work (some of which I myself witnessed), they reported that, though substantial repairs were desirable, the building was very strong, and repair and restoration should prove more than adequate. As an extra precaution, however, I spoke to the concerned Minister at the Centre, Dr. Nurul Hasan who, arranged

for research specialists to send a team on priority to us in Mumbai to investigate and advise us. Their report was also very favourable, but they did stress that, whereas the basic structure of the main building was solid, apart from substantial 'surface' renovation, the drainage system was very defective and must be extensively redone.

Greatly encouraged, I called a formal meeting soon thereafter with the state Chief Engineer and his team of specialists for detailed discussions. I might mention that I was not ignorant of the technical aspects of such engineering, as my father had been a very highly qualified engineer (ultimately, the Chief Engineer of the erstwhile State of Hyderabad) and I remembered well that he often stressed certain key areas that should always be kept in mind. Ultimately — and fortunately — the state Chief Engineer and his team of specialists were clearly fully convinced that together with fairly extensive renovation of the exterior, repair of walls and flooring of certain rooms, plus a completely new drainage system would be more than adequate, and, indeed the preferred option. I then asked the Chief Engineer for a formal report with estimates. When this arrived, I was delighted that the final agreed plan would also result in a substantial saving in expenditure — just around Rs. 25 lakh against the earlier one approved in principle with an estimate of Rs. 1.25 crore. This was of special importance as, at that time, Maharashtra was going through a particularly serious period of drought, and austerity was rightly, the order of the day.

The Governor's residence at Raj Bhavan when we first moved in Mumbai, May 1982

The Governor's Residence, after repairs and renovation

The execution of the work was extensive, and everything from the main block, which included our residential rooms, the dining and reception rooms, the kitchen and my offices had to move for several months to the block across the eastern lawns for which the repair, renovation and modification works had

been included in the overall plan. Work started. Happily my wife, who was highly qualified in interior design and skilled, undertook the responsibility of not only overseeing the interior furnishing and decoration but often, very often, was actively involved in the selection of items of key importance. In this, I clearly recall that, to minimise expenditure, she carefully examined all the old items of furniture and some paintings that had been dumped in a store-room; and, after careful selection, had them restored at a fraction of the cost of new construction. Occasionally, when doors had to be replaced she would ask the engineers what their estimate was, and, for alternatives, she would go to what was (and probably still is) called the "Chor Bazaar", and bring back beautifully carved replacements well within the estimates of the engineers. The corridor and entrance hall of the main building still bear eloquent testimony to her involvement. Her work in this specialised field is, I know, remembered with pride and admiration to this day by residents, visitors and staff. My own direct involvement was restricted generally to the exterior appearance, my own office and a few important public rooms and halls. I also firmly insisted that the main building should, by adopting the tiled roof finish, be in complete harmony with the other adjoining buildings of earlier years.

I was also fully involved in the extensive relaying of the main gardens and lawns; suggesting, I remember, before we left, on getting a few domesticated peacocks to add more colour to the lawns seen from my office. Finally, and very specially, I closely monitored the totally new layout and finish of my own office and the adjoining rooms. Whilst restoration and construction were in full swing, dozens of carefully selected coconut palms were also planted on the northwestern edge of the estate.

Since the renovation plan included the repair of the leaking Durbar Hall and the separate magnificent banquet and reception hall building, the amount of work for the engineers was really very extensive. But a stage came when I felt the work was never ending, and set the following 20th January 1983, barely a month and a half away, but just in time for the annual Republic Day Reception, as the target date for completion. To the very great credit of the State Engineers, they met the challenge. Indeed, they performed so

magnificently that I happily persuaded the Chief Minister and his Cabinet to attend the inauguration so that we could, all together, compliment the Chief Engineer and his team. The photographs above illustrate their achievement well, I think.

"Financially, Also, We Saved a Crore of Rupees!!"

What is known at Raj Bhavan as the 'Point Bungalow', the cottage on the western edge of the estate, was of special interest. We were told that from Independence Day the Prime Minister, then Pandit Nehru, during his visits to Mumbai, always preferred to stay at this cottage. Prime Minister Indira Gandhi also, from those early days, had become very attached to it and primarily for sentimental reasons, she once told me, was not at all in favour of any additions/alterations. This building, as a consequence, apart from normal maintenance, was never touched. All along, however, I strongly felt that, keeping in mind the importance of the guests, quite drastic changes were most desirable. For instance, although it was beautifully located, right on the sea, the main rooms (i.e., the drawing room, the dining room and both bedrooms) were so located that none had a view of the sea. To make matters worse, there was not one decent bathroom for the Prime Minister and his/her family.

In '83, therefore, with the State Engineers, I inspected the site, pointed out the shortcomings, discussed possible alternatives in great depth and prepared a plan. As the Prime Minister was scheduled to stay at Raj Bhavan for a brief visit soon thereafter, the engineers, on my instructions, quickly prepared a small model of the cottage incorporating the rather major changes we had in mind. On her arrival, perhaps a fortnight later, just before her departure, I persuaded her to visit my office to have a quick look at the changes I had in mind. I quickly, but, knowing her sentimental attachment, with a degree of concern, I have to admit, explained the main shortcomings of the existing construction, and requested her to consider the rather major modifications I had in mind. To my immense relief, within minutes, she accepted all that we had proposed! In fact, before leaving, she said that after a few months she was due to visit us again and hoped that

the work would be completed by then. This spurred us all to do our best and, quite soon thereafter, when the work was completed to my satisfaction, we immediately informed her office. Sadly, very sadly, indeed, most tragically for the entire country, Fate intervened: weeks later came the terrible news of her assassination. It was, indeed, a shattering blow. However, months later, when Prime Minister Rajiv Gandhi and Sonia came and stayed there, we were all delighted to see how happy they were with the changes.

Development of Extensive Vacant Land at the Pune Raj Bhavan

I think it was in early 1984 when I was at the Pune Raj Bhavan to review the proper utilisation of the very extensive land area that had been allocated to it — and which we really did not need, nor indeed have the resources to look after and keep in a befitting state. As part of it, with the help and willing involvement of the Forest Department, we created a modest animal and bird park with several kinds of almost domesticated deer, rabbits, and beautiful male and female peacocks, all surprisingly — and delightfully — tame and friendly. I had no hesitation in accepting the advice of the Forest Officers involved, to give very controlled access to the park for the public through a relatively small wicket gate, and open for limited hours. Additional security was, of course, also carefully planned and strictly enforced. Soon we added a bird house, again with a variety of almost domesticated pigeons, doves, partridge, quails and other lovely coloured birds, all seemingly at least (!) excited and happy when fed by hand. Very soon, its popularity grew and even distinguished visitors from Delhi, including the President and Vice President, who, I well remember, extended their stay for a day or two to enjoy an early morning visit to the bird house to sit with many birds flocking close to them and waiting to be fed. In fact, occasionally VIPs would specially call from Delhi to say how much they enjoyed their stay because of the park! Apart from being a welcome addition for all at Raj Bhavan, there was an appreciable reduction in the manpower required for the upkeep of a large portion of the estate.

In spite of this, large areas were still unused and vacant, for acceptable maintenance of which much manpower would be necessary. I then decided

to revive an earlier proposal of a former Governor, my father-in-law, the late Nawab Ali Yavar Jung for the Bal Kalyan Sanstha, for building a centre for Therapy and Recreation for the Physically Handicapped. Fully funded by the Social Welfare department, it was decided that the Lady Governor (my wife at that time) would be its President. Bilkees, as always, whenever a responsibility was entrusted to her, became deeply involved in this project and had a team of involved and dedicated citizens on the Board of Management. It proved a great boon as, soon after completion, children of no less than 13 centres for handicapped children began using it regularly. Their joy and happiness was always a heartwarming and enriching experience. Finally, another substantial piece of the estate was made available for the Yashwantrao Chavan Academy of Development Administration, an academy which, I learnt recently, has become a vibrant centre for the training of government officers in different aspects of development administration.

The Chief Minister of Maharashtra, Shri Vasantdada Patil, with Bilkees and Idris

Bhiwandi Communal Riots

In 1983, as is still remembered, tragic communal riots broke out in the Bhiwandi area of Mumbai, taking a tragic and dreadful toll of life and property, particularly of one community. The Prime Minister, Mrs. Gandhi herself, flew down to make a personal assessment of the terrible calamity and the remedial measures needed. I was, of course, with her throughout, and so was my wife for the tour. Particularly comforting to those whose lives had been devastated, I still remember, was the small intimate meeting that Mrs. Gandhi insisted on with a select number of those who had suffered, primarily to draw them out, narrate the events as they saw them, and then suggest remedial measures.

The public sympathy was overwhelming and heartwarming. Mr. J.R.D. Tata, whom we knew well, immediately and spontaneously donated Rs.2 lakh to finance the relief programmes initiated at Raj Bhavan by my wife and under her personal control and management. Many of the wives of the officers and staff of Raj Bhavan immediately volunteered to help, and often worked late into the evening, packing and making kits with food items and household needs procured to help families restart their shattered lives. And then, early on the following morning, every day, for days on end, my wife would go to the riot-affected areas where, in a very organised manner, with the help of the local Collector, they would hand out the kits to each carefully identified devastated family. Altogether, around 6,000 kits were distributed. Another rather unique, and also rather courageous initiative my wife took, soon after the rioting started, was to personally visit the more communally inflamed areas, and, not only try and console them, but also, quietly and gently, reason with them, highlighting always the oneness of all.

My wife then organised the rebuilding of 750 homes for both communities in Bhiwandi which had been destroyed or badly damaged. For this, a Jain group donated the funds, and Muslim architects worked on the project. For rehabilitation with communal harmony, she was able to place Muslim social workers in the affected Hindu areas, and Hindu social workers in affected Muslim areas, with Christians taking up the task of vocational

training. This was a unique and inspiring demonstration, all felt, of what our country stood for; and happily, this proved to be a most promising end to an otherwise shattering tragedy. Once even Mother Teresa asked my wife to take her along when she went to visit the victims at the camps, and there she also offered to care for the affected children.

Thus, ends the above note written by Idris so many years ago.

It revives my own memories of that terrible period. With the help of the local authorities including the Social Welfare Department, we enabled every widow of the Bhiwandi carnage to earn a livelihood in a method of her own choice. I must mention two experiences that illustrate how effective Gandhian actions of inculcating values and non-violence proved to be in different situations, and at crucial times when lives have been at risk in two incidents during the communal riots in Bhiwandi and Mumbai. There was great tension following the wave of killings, arson and destruction of life and property. Shrimati Indira Gandhi who was the Prime Minister, flew down to Mumbai. My husband took her to the rescue camps and to the very disturbed red-light area where there was great tension after a killing. I was also with them. Hordes of women rushed up to Mrs. Gandhi from both sides. She spoke with them for some time and then she and my husband left.

As soon as they left, numerous women on the street rushed at each other, fighting, tearing each other's hair out, screaming and using bad language. Although many of them had been able to get near the PM, there was a large number of others who had not been able to and, thus, resented the others. I immediately contacted Mani Bhavan, a mini museum-cum-Gandhian Centre which was run by Shrimati Usha Mehta, a wonderful former freedom fighter. I requested her to send their group of *bhajan* singers who often sang Gandhiji's favourite *bhajans*, to this disturbed area.

I was there sitting on the ground in solitary splendour at one end of the street in this red-light district when the *bhajan* singers arrived. We all started singing Gandhiji's favourite *bhajans*. One by one, the women and children came from the sides of the street, and joined in. A *sadhu* was going by in a truck, and I waved to him to join us. He came along, and a Muslim from nearby also came and each of them recited a prayer. Then suddenly the Senior Police Traffic Superintendent, a very fine

Sikh officer, came along. He came by to find out what was going on, and was surprised to see me and to see this crowd, sitting and singing when he had been told that they were fighting! He also recited a prayer from the Guru Granth Saheb. The atmosphere changed, the anger dissipated, and those women embraced each other with warmth. There was no more violence in that area thereafter.

Towards the end of the same riots, there were stabbings of one or two persons every evening near a particular railway station, as people walked out of the station. Women who worked at distant factories went and came by train.and were at risk of being killed. It was a volatile area dominated by an aggressive political group. On the other side, was a large building in which many of the widowed wives, daughters and mothers of the riot victims with their children were being given shelter and care. The wailing of the women and children was deeply distressing.

As the train arrived and the women came out, I asked several of them if they felt safe. A large man with the air of an arrogant leader came up and aggressively asked me what I had come for. The crowd gathered, closing in to listen, as our crowds tend to do. He spoke rudely in a threatening and angry voice. I had to cool the atmosphere to avoid violence. Again, Gandhian non-violence worked, as I asked him not to get riled up for no reason, for I could easily have been his grandmother or sister in a previous birth, or he could have been my brother or father, so why get worked up! The crowd broke into laughter. And there is nothing like laughter to diffuse tension. The result was that we talked right there about the violence, the bloodshed and killings, the terrible hurt, sorrow and cruelty, and how we had achieved even the independence of the country, mostly non-violently. Lo and behold, the assembled men and their leader promised that they would not be the aggressors, that they would never in future initiate violence! After that, there were no more killings there.

Most unexpectedly, Idris was made Lieutenant Governor of Goa while concurrently remaining Governor of Maharashtra

When the term of the incumbent at Goa had come to an end, Idris was appointed the Lieutenant Governor of Goa also. Idris found that as Lieutenant Governor of Goa, he was also the final authority there in matters for government approval which extended from the senior-most down to postings of the lowest level of officials. This entailed

enormous amounts of paperwork and of studying files. Idris requested Mrs. Gandhi to appoint a full-time Governor there.

Guard of Honour by the Indian Navy for Idris, on his arrival as Lt. Governor in Goa

Goa is beautiful, not just the beaches and coastal areas, but the old churches and other historic monuments, and has such warm and friendly people. The famous and holy St. Xavier's Church of Bom Jesus drew pilgrims to Goa from far and wide, in addition to endless numbers of tourists. There were charming old Portuguese-type mansions and lovely little homes as well as churches and wayside shrines in numerous places all over. We were sorry to see that the beautiful Se Cathedral, with paintings on the wall depicting the life of St. Theresa, had salt water dampness seeping up the walls. Idris spoke to the Archaeological Department about such matters that needed very special attention by experts, and was able to get expert action initiated right away. However, while preparations for the forthcoming CHOGM Conference were on in Delhi, Goa was to be where the Heads of State would come for a weekend retreat.

Idris was doing his best to activate and coordinate all that was to be done, but he requested Mrs. Gandhi to come for a visit as it would help to finalise and expedite arrangements for the CHOGM Retreat.

Idris receiving the Prime Minister in Goa (as Lt. Governor), 1983, with Rear Admiral Stan Dawson on her right

When Mrs. Gandhi came, she was incredibly meticulous in her detailed observations at each stage. She was staying with us at Cabo Raj Nivas, but her programme was incredibly crowded throughout. She was caring and polite with everyone. In Goa, she was driven endlessly over very dusty roads which must have been very trying, yet she would stop and talk with any of the people who gathered on the roadside to see her and offer her garlands. She addressed large gatherings as arranged by the state authorities, in addition to the meetings for the CHOGM visit. Finally, in the evening, when she came out of her room for our formal dinner, she smilingly remarked on how refreshed she felt after a bath.

The Cabo Raj Nivas was the historic and lovely Governor's residence. Its long driveway through tall trees and open front garden was at a height overlooking the sea. It had been a nunnery at one time, and had a historic little church attached to it

facing the sea. I had been very concerned about the proper conservation of so many very special buildings of this unique state.

Mrs. Gandhi mentioned to me the next day after her breakfast that Idris had stressed the need for Goa to have a full-time Governor again soon. She said that she would like me to stay on in Goa as Governor. I was deeply honoured at her faith in my ability, and said so. But I explained that I would not like to be away from Idris who would be in Mumbai, and, therefore, I would not be able to do so, however much I would have liked the challenge, especially in this most wonderful state. Mr. Alexander also discussed this with me later. It was a very great honour to have been been asked.

As Lieutenant Governor of Goa, Idris continued to get large piles of important files dealing with various urgent matters. It was also necessary and important to go frequently to Goa. The CHOGM arrangements worked out well, and everything was arranged on schedule. It was a great pleasure for both of us to get to know some fine people there. The Chief Minister Shri Pratap Singh Rane really knew the people of his state, and was on first-name terms with those of varied status everywhere. His lovely wife Rani Vijaya Devi was deeply involved in various spheres of social work, and both of them have continued with their concerns and involvements in the state throughout these decades. Although it was lovely to be able to keep going there, it was an involvement that required a full time Lieutenant Governor. Soon, however, Idris' request to the Prime Minister to appoint one bore fruit, and so this additional responsibility ceased, but our love for Goa and our visits there have continued.

CHAPTER

27 Environmental Initiative and Other Involvements

Again I am reproducing excerpts from another of Idris' vast notes, this one on aerial seeding, in which he wrote:

> *Aerial seeding was another project that I undertook in Maharashtra, which I feel, in a limited way at least, was a success. Almost within the year (as far as I remember), Shri Vasantdada Patil, a very senior, respected and politically strong personality, replaced Shri Babasahib Bhonsle as the Chief Minister. In spite of his indifferent health, the newly appointed Chief Minister quickly made his presence felt. He was both cooperative and considerate. I was delighted that we were able to establish a close and comfortable working relationship. In fact, on more than one occasion, when I suggested certain actions or programmes, for example, aerial seeding for reforestation, for his approval/consideration, his spontaneous response was that I should feel free to initiate, without reference to him, any programme that I felt was in the interest of the state, and he would ensure that it had full government support. In fact, at our very first meeting after he came to replace Shri Babasahib Bhonsle, when I suggested that the state may, after detailed examination by specialists, consider aerial seeding for restoration of extensive barren areas which in earlier years had been rich forests, I was delighted to find that in his very first address to the Assembly, he announced that his government had decided to implement such a programme on high priority.*

I must mention that, being especially keen on the preservation/restoration of forests and other natural greenery, I had studied the technique of aerial seeding for afforestation (very successfully practised worldwide) in depth and, as part of several programmes initiated by the central government, had gained considerable first-hand experience. In fact, on one of my early visits to Delhi I had also mentioned this to the then Prime Minister (the late) Shri Rajiv Gandhi, adding that, since the aircraft element would probably prove to be the most expensive, and may discourage some, I would happily convince the Air Force to willingly participate in this national development initiative. To say that the Prime Minister was enthusiastic would be an understatement! In fact, in my presence, he rang up the concerned Ministers and directed that it must be pursued on high priority by the Centre also.

Now, greatly encouraged by the state government's unqualified support, and determined to take full advantage of the forthcoming monsoons, I called several meetings with the Chief Conservator of Forests and his team, at which several of the State Ministers were also happily present. Extensive trials were conducted; trials in which, because of my past experience, I decided to participate in person. Observations and the encouraging results were documented by the department, copies of which were given to me and which, fortunately, I have carefully preserved. Soon after the trials started, the enthusiastic support of the press and the public was also heartwarming. In fact, the donation from the late J.R.D. Tata also covered a rather desolate area at one of his hydro-electric plants near Pune. Some years later, I received a wonderful photograph from one of the more positively minded (!) forest officers, showing a lush bamboo patch that had enveloped the designated area. I also recall that in response to public opinion, reflected to me also by my wife who had earlier visited Elephanta Island personally, and had said that the island had lost all its trees and was dry and barren, we included this area as well. Within the year, this had successfully added much to the greenery of this island in the vicinity of Mumbai.

When I was later appointed Ambassador to France, I fortunately had with me copies of one or two of the Forest Departments reports that were given to me after the initial trials. In France, soon after my arrival, I shared our

experience with the then Secretary General of UNESCO, Mr. M'Bow. The latter was so delighted and felt that it held such promise, that he asked me if he could visit Maharashtra for a personal exposure and discussions. I was sure that this would be welcomed, and the next day spoke to my successor in Maharashtra and later President (the late) Shri Shankar Dayal Sharma, whom I knew well. His immediate response was that Mr. M'Bow would be most welcome and that he would be happy to have him stay at Raj Bhavan as his honoured guest. He, however, added that before we went ahead, he would first check on the current status of the programme. Four days later, he called me to let me know about the latest response of the Chief Conservator of Forests and his staff, who at formal meetings had earlier enthusiastically and unreservedly supported the programme, as they had also done even at my meetings at Raj Bhavan. They had thereafter, the Governor told me, terminated the programme as, in their opinion, it was not worth pursuing!

Incidentally, it is worth noting that, over many decades, the very same method of afforestation has been widely used by many countries, including Australia, New Zealand, Canada, and, deserving special mention, by China, where, on the Internet also it had been reported, that, by this technique, over a million hectares had already been successfully afforested by them. Sadly, we, of course, decided that the proposed visit by Mr. M'Bow should be diplomatically called off.

Understandably, I was shattered by what I can only describe as the 'double-speak' of the leadership of the Forest Department at that time. To be fair, however, I have to say that, when in Paris, I received one or two revealing (but deeply saddening) private personal letters from a couple of the middle-ranking officials of that department. They had been intimately involved in our trials, and had always enthusiastically supported it, and wrote expressing their anguish at being, in a sense, silenced by the leadership of their department as soon as I left.

In conclusion, I feel that I must record that I very strongly continue to feel that, in the national interest, the afforestation by aerial seeding initiative

that we took on in Maharashtra, is of very great importance, not only for the state but also at the national level. A record of the results as contained in the official report of the Forest Department of Maharashtra is also with me. Should, at any time, government at an appropriate level, be interested, I have with me comprehensive material for reference on the key objectives, the technique and its great advantages over current conventional methods of our Forest Department.

Non-Conventional Energy

During my quite frequent visits to the rural areas, it was naturally very distressing to see, first hand, the enormous burden on the women. Particularly distressing always was to see them — very often, old women, bent with age — first locating, then breaking, collecting and carrying enormous loads of firewood for their daily cooking, and finally lighting their choolas and cooking in small, airless smoke-filled rooms, day, after day, after day.

Of course, in a few, very few, villages, there were some government-sponsored biogas plants in the odd home, in the odd village, but clearly they were primarily for the purpose of demonstrations for visitors — or so it certainly seemed to me. To me, it seemed that solar cookers developed by the state held great promise, and I immediately sent for the two or three models that had been developed and successfully tried out, to personally try them out at Raj Bhavan and satisfy myself. Our repeated trials proved most encouraging, except that, for use in the rural farming community (which had been my main focus), some relatively minor design changes were clearly desirable. For instance, though in the existing models, the size and number of internal boxes were well suited for the urban middle class, minor redesign for the poorer rural community was essential. For this, I suggested to Mr. J.R.D. Tata (whom I knew very well) that he might consider taking it up as one more measure of social service. As expected, he readily agreed, provided the government was also committed to its promotion. Additionally, the administration would have to ensure some simple and limited maintenance and repair facilities at selected accessible locations, e.g., PDS shops.

My final task, I felt, was to convince the political leadership at the highest state levels that for the poor of the rural areas it was indeed the "Magic Box"! Towards this end, one afternoon, I invited the Chief Minister and some of his more senior Cabinet Ministers for a lunch at Raj Bhavan, and had a more than adequate but simple meal prepared right in front of them, telling them that that alone would be our lunch for the afternoon! I was delighted and greatly encouraged by their enthusiasm and their undertaking to carry demonstration units on their tours. Sadly, very sadly, I was unable to ensure its implementation as I left Maharashtra for Paris soon thereafter. However, I have decided to include this piece also in my notes as I feel that, together with aerial seeding, it also deserves to be pursued in the interest of the people of the state, both rural and urban.

Incidentally, in this connection, it is, I feel, worth mentioning that when the (late) President Jaywardene of Sri Lanka was on an official visit to our country and stayed with us at Raj Bhavan, he too was thrilled and excited by the device. In fact, so much so, that he told me that he was determined to introduce it in his own country. Seeing his genuine interest and his enthusiasm, I presented him with a few units as specimens. I clearly recall also that, soon after his return, he wrote me a charming letter of thanks, mentioning that he had already "got the ball rolling!"

In my quite frequent meetings with the then Chief Minister Shri Vasantdada Patil, yet another proposal I made to him which he received with encouraging enthusiasm was solar lights in villages which had little or no electricity.

As I was appointed Ambassador to France soon thereafter, I was unable to monitor the steps taken and the progress achieved in either the production and use of the solar cooker or the solar lights programmes. These were programmes which, I was given to understand by the Chief Secretary before I left, had been, on the instructions of the Chief Minister himself, entrusted to the concerned specialist department of the state government.

Photography

Another great interest for Idris' was photography, which has continued since his

childhood. While we were in Mumbai, Idris was persuaded to send in a few of his photographs, but as exhibits only, when a public competition was announced. We were thrilled when he was awarded a prize. Soon thereafter, there was also a full page article with his photographs in *Illustrated Weekly*, which was then an excellent magazine under the illustrious Khushwant Singh. Even in Idris' early years in the Air Force, while he was an Instructor at its Initial Training School, he bought his first movie camera and took movies. He then suggested that as the trainees were not able to see any flying in that place, he could go to a flying station and make a video film of the aircraft flying there, which would be of special interest and use for the trainees. The authorities saw his other films, approved, and then he was sent to Jodhpur where he made two films and brought them back for the cadets to view their flying sequences. We have also enjoyed numerous films that he took wherever he or we as a family, have been. They continue to refresh very precious memories for us.

While we were in Mumbai, a delightful event was when J.R.D. Tata repeated his flight of yesteryears and flew his plane solo (if I remember rightly) again from Karachi to Mumbai. We went to receive him as he landed and then taxied in and got out of the aircraft happily smiling. This was a real achievement at his age. He was such a special person and had always been a very warm and a caring host when he invited us for a meal with him. When we hosted a banquet in his honour at the beautiful banquet hall of our Raj Bhavan, he was like an enthusiastic young man again. At the end of the dinner, Idris spoke admiringly about J.R.D., but then J.R.D. got up and in his speech he spoke about his flight but then went on to shower great praise on Idris and on me, ending by saying that we were the best of all the Governors and their wives that he had observed in his long life! They were kind and very gracious words, and coming from him it was praise worth treasuring.

I Get Involved in the Dharavi Slum

While Idris was very busy with numerous official and social tasks, I also had quite a number of matters to attend to, officially and socially. I became deeply involved in a number of concerns in Mumbai and in Pune. One very special involvement for me became this constant stream of destitute migrants who came from Kerala, Tamil Nadu, Karnataka, Andhra Pradesh, Uttar Pradesh, Bihar and, later, from the northeast too, with almost half more also coming from the villages of Maharashtra.

The slums were constantly growing and almost half the population of Mumbai were slum-dwellers. My efforts were especially also to see that their children and youth were educated, provided with vocational training to earn an honourable livelihood. For the former, I had to get the children out of the clutches of the Dharavi slum mafia, which, fortunately, I was able to do successfully.

Soon after we moved to Mumbai, in 1982, I went to the Dharavi slum. I had been requested by Mrs. Wankhede, the Minister's wife, to come to the only school there, as an Indo-German group headed by a Dr. Ganeshan was checking out children for leprosy. I saw a large man handing out medicines to children who had been identified as having leprosy. When I praised him for this to the lady ward officer, she told me that he was a mafia don and that the first day that she herself had gone there on duty, this particular man, a Dada of the area, had blocked her path at the entrance lane to Dharavi. He had bent down, picked up a stray cat going by, twisted its neck and thrown it at her feet saying, "This is what happens when you come here uninvited." Every street of Dharavi was ruled by a "Dada" or Mafia don in those days, and I saw groups of children brewing illicit alcohol for the mafia Dadas to sell. They were lured out of school, enticed by a salary and also a good mid-day meal. I then arranged for a meeting there with over 90 Dadas of Dharavi, and discussed the future of their children with them. They agreed to their children following a different path. We decided to try and get vocational training for them, and I established the Society for Human and Environmental Development (SHED), with several concerned citizens who agreed to join the Board of Management, in November 1982. By the end of February 1983, we started the first Vocational Training Centre in Dharavi. It was not easy as there were no funds.

In the beginning, we began with immunisation of the children, and other special programmes on hygiene, health and then family planning. The terrible riots at Bhiwandi came about in the early summer of 1983, and then spread through many areas of Mumbai. This became another deep concern, and I can still hear the heart-rending screams and feel the terror of the children and women in the affected areas. They haunt me when I think back about that period. I went to the camps and the homes of those affected almost every day for some time. As my husband has written in his report, I was able to organise supplies and pack and distribute over kits consisting of 21 items of basic household requirements for over 6,000 families. An unexpected

call that I received one day was from Mother Teresa who said that she would like to come with me when I went to the camps of those affected in Bhiwandi. I was deeply touched and honoured. It was a wonderful feeling to be sitting with Mother Teresa. I mentioned to her that I wanted to do so much but had a funds crunch, and did not know how to collect funds for our work for the needy. She said, "You have to forget about your pride and beg"!

The Dharavi slum is almost a mini-India as over 300 migrants from nine different states of the country flock to it every day, so that there are now almost sixteen lakh people said to be staying in this vast sea of humanity. Our projects for the people have continued through these 30 years and have spread, so that SHED now also has projects and centres for welfare work in five slums of Mumbai and in 30 tribal and other villages of Thane District. The society may not have enough funds, but they have wonderful people, including those of the slums and villages who work with their hearts and heads and who understand the ethos of the local people. When we left Mumbai, Squadron Leader Mohan Rao became the first and most dedicated Chief Executive of our society, SHED. Sadly, he passed away over two years ago.

As activities grew, they included 64 youth groups who support our work for special social causes. Every year, at the time of the Ganesh Pooja, these youth groups consisting of youngsters of varied faiths, set up Ganesh mandals and hold workshops with the support of the temple, mosque and church elders. They have open discussions, perform street plays on special themes such as Population Control and Health, AIDS Awareness, National Integration, Citizenship, Anti-corruption Values and Communal Harmony. They support our efforts to understand and spread values, especially of Gandhian truth, peace and non-violence as well as gender equity. Over 120,000 young people have availed of SHED's vocational training in centres in five different slums at Kalina , Jogeshwari, Andheri, Saagbag, including Dharavi and, in our village project, which is provided along with awareness training in values and good citizenship. We helped to cure large numbers of children with leprosy as well as children and adults with T.B. Recently, in some municipal schools, especially for girls, we have had a year's project for hygiene and health awareness. Toilets were built and others renovated in these schools also.

Sunil Dutt presenting Bilkees with an award for social work

President Giani Zail Singh at Raj Bhavan with Idris and Bilkees, Raj Bhavan, Mumbai 1984

Over 48,000 families adopted methods of family planning, and 49 Mahila Mandals have been established which help to keep the peace. Our special Family Counselling Centre has resolved over 35,000 cases of broken families or harassed girls. SHED also helps over a hundred destitute old grannies, who now help us in turn with our children's crèche and our seven Balwadis. SHED personnel have worked on a vast project for AIDS awareness and prevention throughout Dharavi, and recently with 60,000 male migrants and 3,000 female sex workers. We have also taken up similar projects in tribal and rural villages around Sephale in Thane District, where we have got over 2,000 toilets constructed along with the local tribals and villagers, and have got every child into a school. There were also drug addicts that we helped to cure.

There are several Raj Bhavans that the Governor stays at in different parts of Maharashtra. Apart from the lovely Cabo Raj Nivas that we occupied on visits to Goa, there were three Raj Bhavans in Maharashtra. The main residence for the Governor was in Mumbai. The Raj Bhavan in Pune had several buildings in the estate, apart from the Governor's residence. The original very imposing and large main building and its vast wooded grounds had been given for the Pune University. The grounds of the adjacent estate which had the present Raj Bhavan were also extensive, and it had a good garden where the Governor hosted the Independence Day function on 15th August every year. Each year, during the monsoon months, the Governor moved to this Raj Bhavan at Pune, along with staff and the officials as required from June or July to the end of August. This was a time when the Raj Bhavan at Mumbai had to have its windows shuttered, especially on the three sides that were exposed to the heavy waves of the sea, and were lashed by heavy rains. It was not possible to arrange the Independence Day function on the lawns there. There is also a residence at Mahableshwar and a very imposing one in Nagpur where the Governor went to stay when he had to move there for ten to fifteen days during the winter session of the Assembly. It was the only one of the three that was built as a Governor's residence. It was huge and palatial, had a well laid-out garden with enormous grounds and separate houses for the staff.

One felt that it was a great expense to maintain three such huge establishments, but it enabled the Governor and his officials to interact with people of different regions of the state, which was important. As regards the Raj Bhavan at Mumbai, with my training in interior design and décor, and my own interest, and study of furniture and antiques, I was gratified to be able to save on expenses, for funds were hardly

available. I had a look at the old damaged items of furniture which had been put aside for disposal in the Raj Bhavan godowns in Mumbai and Pune, and got many of them repaired, restored and put back into various rooms. They are amongst the beautiful old carved pieces of furniture that continue to be used at Raj Bhavan. The challenge was also of tastefully rearranging the rooms with limited funds.

Idris continued to visit the different regions of Maharashtra. One of our visits to Nagpur was in connection with a function at the Nagpur University. I came across Idris' notes on advice to the students which he had jotted down before addressing them. They were so meaningful that I have added them here as follows:

Some Points of Idris' Address at the Prize Distribution Function of the Nagpur University, Organised by the *Nagpur Times* on 14.11.1984

Inexpensive to Build and Yet Most Valuable

Nothing is more important than the unity and integrity of our nation. India is indivisible. The theme of the contest is "Our India: One Nation, One People" — most appropriate at a time when the nation is passing through a great ordeal, when disruptive forces are once again conspiring to destroy the unity and integrity of the country. It is a coincidence but a happy coincidence that the prize distribution is taking place on the birth anniversary of one of the noblest sons of India, Pandit Jawaharlal Nehru, who throughout his life steadfastly championed the cause of national unity. Speaking at Bangalore in 1955, Pandit Jawaharlal Nehru said: "We should not become parochial, narrow-minded, provincial, communal and caste-minded, because we have a great mission to perform. Let us, the citizens of the Republic of India, stand up straight, with straight backs, and look up at the skies, keeping our feet firmly planted on the ground and bring about this synthesis, this integration of the Indian people an emotional integration of the Indian people so that we might be welded into one, and made into one strong national unit, maintaining at the same time, all our wonderful diversity".

Years later, speaking in the same vein his illustrious daughter Shrimati Indira Gandhi said:

"It has always been our endeavour to build bridges, not only across rivers but also in the hearts of the people. We have in India people of different communities

following different religions, speaking different languages and living in different provinces. We wish to build a bridge of this kind between them, a bridge which may not be apparent. What will be apparent is the spirit of friendship, brotherhood, love and cooperation. These are bridges we seek to build."

Our Rich Cultural Heritage

We are inheritors of a rich cultural heritage. Part of this heritage is the tradition of our people, people professing different faiths, speaking different languages, belonging to different regions and ethnic groups, living together in amity and harmony. That is strength. But, at times, this unique edifice built by years of patient labour is sought to be weakened by the misguided actions of a few whose outlook gets distorted by communal, regional, sectarian and linguistic passions.

Example of Armed Forces

In the armed forces, people belonging to different parts of the country, speaking different languages and following different traditions and religions come together and live as one family. This gives you the feeling that you are Indians first, Indians last, Indians always. This example of national integration is worthy of emulation by everyone in the country. A nation divided cannot become strong or prosperous: a nation which is divided within itself can hardly expect to become strong or prosperous. In the ultimate analysis, the essence of national integration is the ability of everyone to intellectually as well as emotionally bracket himself with everybody else in the country, in the hopes and aspirations of his neighbour and in his willingness to contribute to the well-being of the country.

Communal Harmony Essential for the Unity and Strength of India

We must by all possible means strengthen the belief that India is one country and Indians are one people. We are wedded to the ideal of a secular nation. It is enshrined as a fundamental right in our Constitution that all persons are equally entitled to freedom of conscience and the right to profess freely, practise and propagate religion. No religion teaches hatred. Every religion teaches us the virtues of compassion, love and service. Yet we find that in

the name of religion, many atrocities are committed. Only a person who understands and follows his religion in depth, can have reverence for his own religion and also show the highest respect for every other religion.

Idris narrates how his father used to assemble all the children daily and read to them from scriptures of his own and also various other faiths/religions.

It is respect for all religions and mutual respect for the different religions that constitutes the essence of secularism. Acceptance of secularism implies that we respect each other's faith, tolerate the differences in their beliefs and accept the fundamental fraternity of all human beings. Our economic growth, political stability, social cohesion and national interests, all demand that we accept secularism and practise it as a way of life.

There is today a great need to unite people from different parts of our country, people of different faiths, of different castes and creeds to join hands, to work together. There is great need to look at the future of our country on a wide horizon. And in this great endeavour, in this challenging task, our educational institutions can and must play a central role. Only their commitment to these ideals can strengthen the forces of national integration, can help strengthen the unity of the country. There is also need to reorient our education. It should aim at equipping students with an intimate knowledge of the different aspects of this country, including the events which led to freedom, encouraging all studies and activities which lead to greater understanding between communities and states, thus, fostering a feeling for national unity and creating a feeling that the country and its resources belong to the citizens who thereby acquire certain rights and privileges along with corresponding duties and responsibilities.

The Role of the Press

In promoting the concept of "One Nation, One People", the press can and should play a positive role. In a vast country like ours, at some place or the other, unfortunate incidents do keep occurring. If these incidents are highlighted and exaggerated in the press, it would only aggravate the situation and may even spread hatred and violence. On the other hand, many noble deeds of tolerance, compassion and help and solace given by members of one community to the

members of other communities in difficult situations go unreported. If the press highlights these instances, it will help promote unity and communal harmony.

Our Responsibility

The great responsibility of uniting India and strengthening it lies with the people. We owe it to our ancestors and more so to the generations that succeed us, to leave this country more united and more prosperous. Our concern should be what kind of India we shall bequeath to future generations — a strong, harmonious self-confident India or a feuding, weak India. When we talk of national integration we should not take it as a mere phrase but should be fully conscious of the burden and task that history has placed on our shoulders.

Build Indians

The great Italian patriot Garibaldi, after he completed the task of uniting Italy, said: "Now that we have built Italy, let us proceed to build Italians". Let us also now direct our energies to building Indians who alone will sustain the greatness of India by working together as one nation and one people.

Our nephew Murad Yar Khan and Asgar at Raj Bhavan in Pune

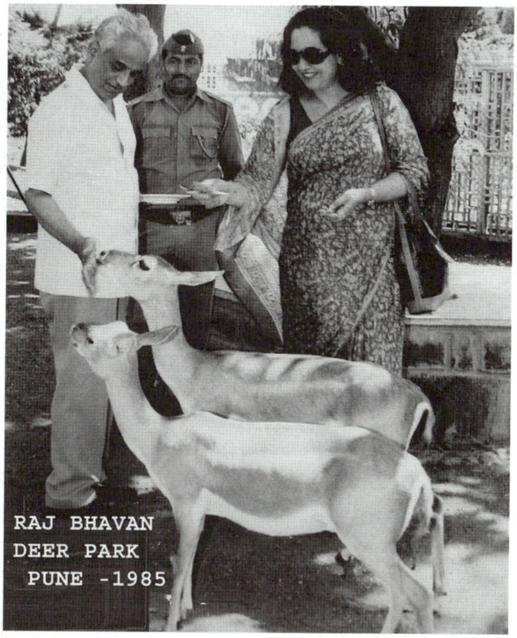

Idris and Bilkees feeding the deer at the Deer Park at Raj Bhavan, Pune

Idris and Bilkees at a Raj Bhavan formal function, Mumbai, 1984

In Pune, while Idris was busy with numerous important matters, I got very involved with establishing the Bal Kalyan Sanstha, a centre for recreation and therapy for handicapped children, next to the Pune Raj Bhavan, which Idris has referred to earlier. When the Bal Kalyan Sanstha was built and activated, it had a swimming pool, physiotherapy facility, library and teachers for dance, music, handicrafts, and so on. Right away, there were large numbers of children who came from 13 different centres, for the handicapped would come to interact with each other while using it. Each day, children with two different kinds of handicap would come there. The blind and those who were crippled or were polio-affected for instance, interacted well together. It has been a great success with very fine volunteers, including Mr. Pratap Pawar, whose guidance, sound advice and involvement were invaluable and a great help, in addition to those on the payroll. The inaugural function and opening of the Pune Bal Kalyan Sanstha's recreational centre for handicapped children was done by Idris as Governor, with the following inspiring speech on Children's Day, 14th November 1983:

> *Madam President Bilkees Latif, distinguished guests, friends, first of all, I would like to say that I was quite delighted to accept the invitation, extended by you Madam President, to associate myself with this afternoon's function. I am happy to be here for more reasons than one, not the least being my close association with you over all these years; the admiration I have developed for your capacity, your dedication and your deep devotion to the welfare of the less fortunate members of our society. Your work in Dharavi and several other large slums of Mumbai and in numerous villages beyond, is an outstanding example. I cannot help but pay public tribute to this aspect of your life, my admiration for all that you have done during the years that you and I were with the Air Force, your work for the welfare of the families of Air Force personnel, Servicemen and civilians alike. This included sharing their happiness and their achievements, but sharing even more so their sorrows and, to the utmost of your capacity, trying to solve their problems and helping them to lead more productive lives.*
>
> *On this occasion, I would also like to pay public tribute to our dynamic Commissioner Mr. Sidhu and his team of officials, without whose deep involvement and Herculean efforts I do not think it would have been possible to complete the construction and equipment of this Kendra in such a short*

space of time. In this work, of course, I realise that there have been many others who have also come forward to lend a helping hand and I would like to take this opportunity to express my deep sense of admiration for the meaningful contribution that they have also made.

The establishment of the Bal Kalyan Sansthan represents, I believe, the expression of a deep and genuine anxiety for the less fortunate members of our society, an anxiety which happily I find, at least in Maharashtra, is frequently translated into action.

Some years ago, we celebrated, under the auspices of UNESCO, the International Year of the Handicapped; which was also celebrated the world over, and the International Year of the Child. These events, I do believe, helped to give concrete shape to many positive measures for their welfare. But we all know that in spite of all that has and is being done, in our country and around the world, much remains to be done. We have, as we all know, many, many miles to go. But the distance we have to travel must not overwhelm us, to make us feel that the task is too great — that the distance to travel is too much. I have no doubt that if the government and the public work hand in hand, as, indeed, is the case of this institution, meaningful progress can be made.

I think all of us who have assembled here are happy that you, Madam President Bikees Latif, and the other members of your team, have chosen to inaugurate this institution on Children's Day, as a tribute, so to speak, to Pandit Nehru, whose concern and affection for children in general and the deprived and handicapped in particular, has become a source of inspiration to the people of our country.

Maharashtra, I must say, is fortunate in having so many institutions which are engaged in social work; so many who are always willing to lend a helping hand. What should be the important plank on which should rest the work of such societies and institutions? It certainly should not be just charity and it most certainly should not be pity. Mind you, these are by no means unworthy sentiments, but I think all of us know from personal experience that these sentiments, however good they may be, are to most recipients, only humiliating.

In every society all over the world, economic productivity is amongst the most important criteria of achievement; for the handicapped, it goes further: economic productivity provides self-respect and a sense of pride. It is only then that the handicapped can lead a life of dignity, of self-respect and of optimism, each so vital for life to be worthwhile. The word 'optimism' reminds me of the story of the engine driver who met with an accident and lost a leg. When accosted by a friend who asked him what good he saw in this tragedy, the engine driver cheerfully replied, "I now have to polish only one shoe". It is this kind of robust optimism that any programme of rehabilitation should aim to instill in the minds of the disabled, so that they are able to face life and the world with renewed confidence. This is a formidable challenge, but a challenge worthy of the many dedicated and devoted men and women who are working in this sensitive but satisfying field.

There is much suffering in the world today, suffering caused by want, by poverty and by ill-health. It is not possible for us to eliminate suffering from the face of the earth; but it is certainly possible and well within our power to reduce the impact of deprivation and suffering. We receive so much from society each day. Is it not our duty to return this debt — at least, in a small measure, by way of, perhaps, some small service to others who are less fortunate than ourselves? Let us, each one of us, resolve to use what we have, be it our position, our wealth, our education or our talent, in the service of our fellow beings. This is perhaps the best prayer that we can offer at the altar of God.

Having spent around four decades in the Air Force, I am, on such occasions, reminded of an almost legendary figure of World War II, an outstanding fighter pilot by the name of Douglas Bader who, at an early stage, lost his leg in a flying accident; but when World War II came, he was determined to climb back into the cockpit. But medical rules, of course, could not possibly allow him to do so. He was determined, however, and knocked on every door of officialdom till finally he was allowed to fly again, despite his amputated leg. But courage and determination to overcome a disability or disadvantage is something that cannot grow merely of its own accord. It needs understanding, encouragement and love and also specialised institutional care. And this is what I know this institution will try its utmost

to do. It is a formidable challenge, but a challenge which, I am confident, none who are associated with it will shirk. And in your endeavours, I wish you every success in the months and years ahead.

There were two very interesting but challenging assignments for me during this period. One was when the Prime Minister put me in charge of cocoordinating the programmes of the First Ladies, the wives of the Presidents and rulers of countries who were attending the NAM Conference in Delhi when the Heads of State came there from many countries. At times, it was quite a challenge as there was a large number and some VVIP ladies had their own individual shopping to do and other visits that were apart from the official programmes. However, they were very charming ladies and all went well.

The other challenging but interesting assignment which the Prime Minister asked me to take on was of cocoordinating and seeing to the visit to Hyderabad of Her Majesty Queen Elizabeth II of Great Britain, who came to India in 1983 for the CHOGM Conference. I found the Queen to be so very gracious.

My book, *Her India*, a biography of my mother and her life in Hyderabad from 1928 onwards and as a devotee of Sri Aurobindo Ghosh, was published by Arnold Heinneman. I had worked on it for several years and had done a great deal of research into that period of Hyderabad. I had been given some interesting material on the "Police Action" after which Hyderabad became a part of an independent India. The book was released by the distinguished and great writer, Mulk Raj Anand in Mumbai and by our very fine Vice President Mr. Hidayatullah in Delhi. It was written

Bilkees, Mumbai, 1985

about with serialised excerpts in *The Times of India* and extensively in *The Telegraph* in Kolkata, as well as by Ralph Bultgens in New York and others.

I was nominated by the Foreign Minister to represent India at a United Nations session of three days on the Freedom of Namibia, which was being held in Singapore in May 1985.

Idris often met students groups of different universities, which both he and the students found productive. He had a very caring and warm feelng for them to which they responded, to Idris' delight with disarming frankness, and helped him enormously in identifying problems that needed urgent remedial action.

Having been so very involved in resolving the tragic delays in the students receiving the results of their examinations, Idris welcomed an opportunity to deliver the following Convocation Address about two months before we left Mumbai on Tuesday, 12th February 1985.

Convocation Address by Governor of Maharashtra and Chancellor, University of Bombay, Air Chief Marshal I. H. Latif (Retd).

Mr. Vice-Chancellor, members of the Senate and of the Executive Council, distinguished guests and my young friends, first of all, may I say that I consider it a wonderful opportunity and a source of genuine pleasure to be with so many young people on the threshold of their careers — representing as they do, the hopes and ambitions of so many parents, of so many families, who, along with the distinguished members of the teaching staff, have shaped their success and prepared them for the challenges that lie ahead. And to those young men and women who have laboured hard and justified the faith reposed in them, I offer my heartiest congratulations and good wishes.

To be products of Indian education, education which has an ancient tradition, originating from perhaps the earliest years of civilisation, and to have the opportunity to merge one's talents and energy into the national effort of a free society and a free country, is a privilege that must be valued and cherished. It is a privilege that you enjoy from today, but with it you also assume special responsibilities as key elements in the pace of the progress

of our people. In every country, it is the system of education that can and must liberate the power and determine the direction of national development. In our country, expansion in the field of higher education has, as we all know, already reached gigantic proportions. The number of universities in the country has, since the early years of Independence, multiplied five-fold: from about 30 in 1950 to about 140 by now. In 20 years, colleges have gone up from nearly 1,000 to nearly 5,000; while student enrolment in the same period has increased thirty-fold from 12 lakhs to 360 lakhs. Expenditure on university education has also risen: from about Rs 15 crore in the first Five-Year Plan to no less than Rs 500 crore in the Sixth Plan. And the total expenditure on all forms and levels of education has gone up from about Rs 150 crore to over Rs 2,500 crore in the Sixth Plan period.

The generative effect of the spread of education is, I believe, well reflected in our performance in other fields of national endeavour. Let us see the impact on some key areas: the birth rate per thousand of population has already dropped by nearly 20% from 42 in the mid-1950s to 34 in the early 1980s, no mean achievement, especially in a mainly rural and tradition-bound society. Life expectancy has also risen from around 30 in the first few years of Independence to well over 50 today; and, simultaneously, the death rate per thousand of the population has dropped from about 30 in 1947 to around 10 in 1981. Witness also the performance in the main sector of national activity, namely, agriculture: in our vast rural areas where, as Mahatma Gandhi said, "the heart of India beats"; production of food grains has taken a gigantic leap from a mere 50 million tonnes in the early 50s to over 150 million tonnes this year. Student enrolment in the sector of science and technology has multiplied more than ten-fold, from 3,000 when Independence came, to well over 30,000 today. And India is now ranked third in the world, next only to the US and the USSR in terms of mobilised scientific and technical manpower. Small wonder then that in the field of industrial development also, our country now ranks eighth in the world in terms of installed capacity, range of production and sophistication of manufacture. With such a reservoir of talent available for energising industry, agriculture, commerce and a myriad other fields, we can well gauge the

impact that expansion of education has had for releasing and activating the power within the people of the country for growth and prosperity.

Recognising the correlation between education and national development, and also the need to do all we can in the coming years to generate an Indian renaissance, it would be desirable to review as objectively and dispassionately as we can the strengths and weaknesses of our system. I would like to highlight a few areas which I consider of importance.

First of all, education must mean the cultivation of both the intellect and the personality or character of the individual. In this process, it must promote the development of individuality — an individuality that also keeps in focus and remains in tune with the needs and aspirations of the people of the country. In the great thrust of an ever-expanding educational infrastructure and the plethora of universities, colleges, schools and institutes, the many lakhs of teachers and hundreds of lakhs of students, injection of information into young minds has been taking place on a massive scale. But character training and personality development has not, in my view, received the attention it deserves; indeed, as a properly developed system it is, I believe, almost totally absent. It is, therefore, precisely in this specific area that a drastic change in policy and priorities is needed. One of the foremost aims of our education policy at all levels should now be to provide a strong framework for individual personality development and through it, to shape and strengthen our national ethos. Intellect without character is like a body without a soul — or, if I may use a more poetic phrase — a plant that grows but never flowers. We must recognise that this task represents an awesome challenge to educationists. But it will be, in my view, the litmus test of the true worth of the insights and outputs of all of you who are our specialists and our experts.

Character training, I cannot help feeling is, to some extent at least, a matter of organisation rather than instruction; and also of example rather than of precept. Precepts, of course, have their utility, for instance, in establishing a set of values. Far more difficult, but far more valuable, is the creation of the right environment which helps and encourages young minds to resolve personal and moral issues independently. Those who have, apparently,

been well brought up at home may behave well and work well to please their parents or their teachers; but this can never be a sound enough basis for conduct. What is also needed is a sense of responsibility to oneself, the determination to master one's emotions and physical faculties for the pursuit of excellence and the evolution of diamond-hard criteria for self-judgment. An essential part of the development of a sense of personal responsibility, equally, is pride in one's country, a respect for its traditions and a faith in its future — a faith that must stimulate lives of service and not just self-service.

This is by no means as difficult a task as it may seem. It can be accomplished; it has been accomplished: the Battle of Waterloo, it is said, was won on the playing fields of Eton. Sports and participation in games are invaluable elements for character development. We all know this; but have we done all that is possible to give this activity the important place it deserves in the curricula of our schools and colleges? Sound judgment, tenacity and cool temper, indifference to bodily discomfort, team spirit, and leadership; all these can be nurtured through sports. The pleasurable glow of well-being that accompanies physical fitness, the quiet confidence that comes with the ability to attack and defend fiercely but without ill-will, the exhilaration that follows risks taken and surmounted, all are invaluable experiences for building of character and the development of personality.

Social service and the N.C.C. are two other organisational areas which can, I most strongly feel, play a major role in this respect. They stimulate and strengthen a sense of responsibility, a sense of discipline, a sense of team spirit and again, no less important, a spirit of service — service before self. The desire to serve others must surely be one of the main springs of thought and action if individuals are to make a meaningful contribution to the strengthening of our society and the nation. In Bombay, members of the academic community are specially blessed — with a vast rural hinterland which can absorb huge scales of effort and, in turn, reward participants handsomely in their growth and in their training. We must mobilise these resources. *Social service and the N.C.C. are two organisational areas which can, I most strongly feel, play a major role in this respect. They stimulate*

and strengthen a sense of responsibility, a sense of discipline, a sense of team spirit and again, no less important, a spirit of service, or service before self. The desire to serve others must surely be one of the main springs of thought and action if individuals are to make a meaningful contribution to the strengthening of our society and the nation.

I must touch upon another important aspect of education. There is a tendency, an understandable tendency, in our institutions to fight shy of developing in students a mature attitude vis-à-vis religion. Perhaps this is so because people are over-sensitive to the issues involved and are, therefore, hesitant to reach out and come to grips with this important element in character development. We must not put a negative meaning into secularism — regardless of what the dictionary meaning might lead one to think. One defines it as: sceptical of religions' truth or opposed to religious education. Such a negative approach would deny us a powerful instrument for promoting moral values. The founding fathers of our Constitution surely gave 'secular' a more positive meaning — not the denial of a place for religion in national life but the acceptance of all religions as equal. In any case, there is no question that, especially in a country such as ours, a civilised attitude on religion is a rudimentary requirement for individual development, and undue timidity, in my view, amounts to neglecting an important area of responsibility.

Some stimulus must be organised through our educational institutions to generate and shape the right attitude amongst the youth in this regard and make them better citizens than many of us have been. Actually, only a very simple idea need be stressed: that in each human being there is an element of the divine, and it is the responsibility of each person to develop it in himself or herself and also recognise it in others. This is the essence of different streams of religious thought, and by emphasising this central inspirational idea, a secular and civilised attitude can be promoted and, with it, a more harmonious development of individual personality. Acquaintance with prayers of different religions and hymns and songs of various faiths could also, surely, help our youth recognise the intrinsic purity, beauty and practical usefulness of different religious thoughts.

CHAPTER

28 Ambassador Plenipotentiary

In April, 1985, Idris was posted to France as the Ambassador Plenipotentiary with the rank of Minister of State. He was also asked to represent India at UNESCO, the headquarters of which was also in Paris. This was after the current Indian Ambassador to UNESCO, 'Tikki' Kaul, had completed his term there. However, Idris said that each post (i.e., that of both Indian Ambassador to France and Ambassador to UNESCO) deserved a full-time representative and suggested that a suitable

representative could be sent accordingly for the UNESCO post. This was accepted. For Idris, as he often says, his assignment in France, as in Maharashtra and, indeed, his 40 years in the Air Force, was also a great education and no less emotionally rewarding.

It was early in the evening when we arrived in Paris and were driven directly to the magnificent mansion, bought in Paris many years ago by our government as the residence of the Ambassadors of India. As we entered the foyer, we felt wonderfully welcomed as a huge basket of the most gorgeous bouquet of flowers that had been sent for us by His Highness the Agha Khan was awaiting us — such a thoughtful and gracious gesture.

In Paris, for Idris, not knowing a word of French and with barely a month and a half left to finalise all arrangements for the inauguration of the very first Festival of India which our Prime Minister and the President of France, M. Mitterand himself, breaking protocol, were to inaugurate, was quite a challenge. However, Idris was very touched by the consideration of the French, as he found that on every occasion, even at social functions, groups who were conversing in French would switch over to English if Idris joined them.

Idris' formal presentation of credentials to the President was, very thoughtfully, specially arranged by the French authorities on high priority, within days of our arrival. Once the President gave a date and the formal ceremony of presenting his credentials was over, Idris decided to formally call not only on the French Prime Minister (M. Fabius), but on several of the more important Ministers, including those concerned with the various official programmes to be organised for Prime Minister Rajiv Gandhi's visit. The real challenge, of course, was to get to know them sufficiently well without being able to speak a word of French! But Idris often mentioned to me that it was really a very heartwarming experience as not one showed even surprise, leave alone resentment at this. In fact, most of them, to put him at ease, most graciously and happily spoke in English and each, without exception, repeatedly assured him that they would be most happy, should ever the need arise, to extend a helping hand.

Throughout our three years there, the French remained most helpful and caring in all that we were involved with. Typical of their warmth was Mr. Jean Riboud, a leading and greatly loved and respected industrialist and a close friend of the French

President, who had been appointed as head of the Festival from the French side. Unfortunately, he was seriously ill and in hospital when we arrived in Paris. Very soon after our arrival, when Idris went to see him in hospital, he was greeted with a warm smile. Then, asking Idris to come close, Jean whispered in his ear, "I am deeply saddened that I cannot be by your side to help you, but please remember that the President is a very close friend of mine; remember also that, day and night, I am as close to you as the telephone". These were unforgettable words for Idris, and indicative of the sentiments and the motivation of the cultured and wonderful people of France. This attitude of support was a pillar of strength to us both throughout the challenging and memorable years that we were there. Soon we found that Jean Riboud's wife, Krishna, a direct descendant of Rabindranath Tagore, was also a fine, lovable lady. At that time, she was very involved in setting up her museum of the rare Indian textiles which she had collected.

Idris with French President Mitterand, after presenting his credentials in Paris, May 1985

As the inauguration of the Festival of India was barely a month away, Idris immediately got fully into all that had to be done, and as for the programmes for the Festival, he found that there were some excellent people both in France and in India

The President and Mme. Mitterand's dinner in honour of the P.M. and Mrs. Gandhi

Prime Minister and Mrs. Gandhi with President and Mme Mitterand, Paris, 1985

Reception in honour of President Mitterand at the Indian Ambassador's Residence

Prime Minister Rajiv Gandhi and Smt. Sonia Gandhi
at the Festival of India Reception at our Embassy Residence, Paris, June 1985

At the Paris Air Show, June 1985

who could be counted on to produce perfection. At the Embassy, for the period of our Prime Minister's visit (which, of course, included the opening events of the Festival), Idris had a series of conferences and set up a Control Centre at the Embassy for arrangements. He then allotted specific responsibilities to the more senior officers. At the Embassy, the senior-most Foreign Service Officer, Dalip Mehta, as Minister under Idris, was a pillar of strength; as was Sharat Sabharwal under him, as the head of the administration. In fact, Idris often says to me even now, that Sharat is one of the most outstanding Foreign Service officers he has had the good fortune to work with. Air Commodore Tarlochan Singh, who was the Air Attaché, was put in charge of all arrangements, in coordination with the French authorities, for the Prime Minister's arrival and departure at the airport, including the ceremonial reception with its splendid Guard of Honour. Similarly, the Naval Attaché was asked to organise and be in overall charge of a round-the-clock, 24-hour Control and Coordination Centre at the Embassy for all events connected with the Prime Minister's visit. Similarly, the other senior officers of the Embassy were also given clearly defined responsibilities and personally briefed by Idris on their specific tasks. And, finally, Idris said, that at his final normal weekly conference, a week before the Prime Minister's arrival,

he had a comprehensive briefing. Then, finally, as was his normal principle of man-management, he also gave each an opportunity to identify any areas that, in his/her view, needed further refinement.

The visit of Prime Minister Rajiv Gandhi and his wife Sonia in June 1985 was an unqualified success in every way. This, very largely, both Idris and I felt, was because of their warm personalities as well as the deep interest and warmth of the French towards them at every level, from President Mitterrand downwards.

For the inauguration of the Festival of India, the resplendent French Mounted Cavalry with soldiers on horseback were lined up at the foot of the Eiffel Tower, as Rajiv Gandhi and President Mitterrand arrived. The two dignitaries were escorted and taken up to the first floor of the Eiffel Tower, from where they formally announced the commencement of the Festival of India, the L'Anee de l'Inde.

Festival of India, Inaugural Day, Place de la Concorde, Paris, 1985

At this inaugural function, held high up on the Eiffel Tower, Prime Minister Rajiv Gandhi first presented President Mitterrand a jar containing the holy water from the Ganges river. Rajiv then presented the baby elephant named Kaveri, which had been specially brought from India, he said, for the people of France. Kaveri, incidentally, had come up in a special lift, and was seen swallowing

bananas and looking around without — seemingly, at least — being at all disturbed or uncomfortable. Elsewhere, flowers were also scattered over the river Seine by 10 Indian girls as 10 others poured water from the Ganges into the Seine at the time of the inauguration. Looking down from the Eiffel Tower, one saw the vast spectacle spread through the expanse of the Trocadero, of over 200 colourful folk dancers and other Indian performers whirling away in colourful costumes and the fluttering Indian flags. Even camels and elephants were present and colourful Indian kites floated in the air. There were stalls of exotic Indian dishes: fresh *dosas*, *kababs* and *biryani* laid out all over. Adding to the spectacle were over 200,000 Parisians, many of whom were joining in with the dancers. Far above, on the Eiffel Tower, inaugurating the L'Anee de L'Inde with President Mitterrand, Rajiv said, "Paris has everything. But today we bring you India."

Festival of India Inaugural Day at the Place de la Concorde, Paris, 1985

From a report in *Blitz* on a special issue, dated 8th November 1986, I have included the following excerpts:

Looking back to the year-long "Festival of India" in Paris, four months after it concluded in June this year, a highly encouraging result that is immediately perceptible is the spontaneous response the Festival evoked from the French

public that, for the first time, helped to project India's rich cultural heritage in the correct perspective to the people of France.

And, in a large measure, the success of the Festival can be attributed to the pioneering efforts of India's Ambassador in France, Idris Latif, and his beautiful consort Bilkees, ably assisted by Indian and French government officials. It was a rare and unique gesture on the part of the French government that marked the inauguration of the Festival, when President Mitterrand joined Prime Minister Rajiv Gandhi to declare it open on June 7 last year.

The French Prime Minister, Jacques Chirac, who is also the Mayor of Paris, demonstrated his friendship for India by making available the prestigious Trocadero area for the Festival, blocking all vehicular traffic during the two days of its inaugural ceremonies and flying the flags of India and France atop the Eiffel Tower, where no foreign flag had been permitted to fly in the past! Also, again, only once before, Edith Piaff had sung from the Eiffel Tower. But during the opening of the Festival, a Rajasthani singer from India was allowed to sing from the Tower.

Spontaneous Mini-Festivals

The French government contributed 25 million francs for the success of the Festival, while a culturally involved and dedicated team of their experts worked all the year-round to help arrange seminars, exhibitions and music and dance programmes of Indian ethos, a distinct outcome of all of which has been that French tourists going to India increased by 20 percent. President Mitterrand also arranged the closing ceremony of the Festival at the Elysee Palace, his official residence, when India's Vice-President R. Venkataraman was present. The main Festival in Paris was followed by mini-festivals in several large and small towns of France.

Inspired by this, the Netherlands Embassy also put on display its own collection of very rare Indian miniatures at its cultural centre. At the international music festival in Montpellier, an all-night session of Indian classical music was held, from 9 pm to 7 am, in an old Abbey outside the town, The response from music-lovers was such that the hall was packed beyond its capacity

At Bismillah Khan's concert during the Festival of India in Paris, with him

With Smita Patil at an art exhibition, Festival of India, Paris

M.F. Hussain in Paris at an Embassy function

of 600, to hear the sitar maestro Ravi Shankar, the sarod wizard Ali Akbar Khan and the ace percussionist Allah Rakha. When it finally ended at 7:15 in the morning, the applause from the audience was truly heartwarming — the three musicians received a standing ovation which continued all the way till the artistes actually got into their car to leave for the airport.

Toulouse: A two-month programme that included visits and training by a Madhubani master artist and also a Kalamkari expert from Andhra Pradesh. At the opening of the festival, the Ambassador Idris Latif and Begum Bilkees Latif were received by local hosts with traditional Indian garlands. Similarly, in Grenoble, Alsace, Cannes and other regions, it was often the local French who wanted to have a festival of India which included a series of concerts too, and then they were fully involved in the organising of these.

At the request of several organisations, and also for the Organisation of Economic Cooperation and Development, Begum Latif arranged an exhibition of Indian costumes there and at the Indian Ambassador's official residence that was praised by the prestigious Vogue magazine for its rich and colourful variety that included female sarees, antique achkans, dupattas and brocades.

Another highlight of the "Festival of India" was the staging of an exhibition of centuries-old Indian miniatures by the Musee Guimet, and the Bibliotheque National — something never exhibited before.

Films with French Cooperation

The French Education Department also sponsored an exhibition on the children of India, their dreams and experiences — arranged by a teacher, Nadia Goralsky, who had lived in Goa recently, and organised an exchange of thoughts between a group of children of a school there and one in Paris, through the Alliance Francaise. It is now proposed to stage a similar cultural exchange between the children of France and the children of India at a school in Bombay. (The exchange with the pupils of the Paris school enabled them to know all about Indian mythology, to the extent that they could draw pictures of Indian gods and mythological heroes!) Films on India with French cooperation are being worked out. An eminent French film producer working

on three projects is E. Pope, the husband of the Turkish Princess Niloufer, who used to stay in Hyderabad years ago, and Japanese designer Issey Miyake's models depicted Indian fabrics at a textiles exhibition.

Determined efforts are also being made at the Indian Embassy in Paris to increase economic, scientific and technological cooperation between India and France, helped by the wide interest generated among the French by the "Festival of India".

Incidentally, in a general poll in France a year before Mrs. Indira Gandhi died, she was declared the Most Admired Woman. That popularity and admiration is extended to Rajlv and Sonia Gandhi and is a reflection of the French feelings towards India. Notably, at the Annual Shoot arranged by the President at one of his estates, Rambouillet, several Ambassadors much senior to India's Idris and Bilkees Latif were always invited. Notably, the Indian Ambassador was always given the place of honour next to Monsieur Grosouvres, the President's representative who would , invariably, host the the (pheasant) shoot.

Air India has been flying full between Indian and French airports since the Festival, justifiably winning an award in India for its excellent performance in France. The famous French designer Pierre Cardin is now ordering shirts made in India for export elsewhere.

This then is the opening of a new era of friendship and cooperation between the people of India and the people of France, all helped by the spectacular Festival of India.

As reported by *Blitz*, the French had graciously agreed for us to have the Indian flag draped from the top of the Tour Eiffel, right down the full length of it from outside. We were told that this was the only occasion, ever, on which a foreign flag was draped on the Eiffel Tower. On France's second TV channel, Antenne-2, the inaugural ceremony and the spectacle of the dancers in varied Indian colourful costumes performing all over was covered live at prime time. On TF-1 also they showed M.S. Subbulakshmi singing that evening. Programmes on India continued to spread all over France right through that year.

As reported in *India Today* about the Mela, "With the welcome arches in the background complex of terraces and stairways embellished with multi-coloured water-jets and fountains, the Mela was gathering momentum. The richly embroidered bamboo arches were aglow with lights as hundreds milled around the puppeteers from Calcutta, acrobats, jugglers, magicians and strolling singers. The largest crowds were seen in front of the stunning Panthi dancers from Madhya Pradesh doing their whirling human pyramid, Kerala's spellbinding Kalaripayattu and the Panchavadyam percussionists, and Teejan Bai's regal rendition of the Mahabharata. As darkness fell and the dazzling fireworks began, the crowd participation became livelier. Befitting the French tradition, the longest queues were in front of the food stalls".

The Festival of India generated enormous interest and local involvement all over France. A great event was Peter Brook's play, "The Mahabharata". It was viewed by more than 100,000 people who came to see it in its most dramatic location among the stone quarries beyond Avignon, the medieval town in the south of France which is famous for the theatrical productions at its annual international festivals. Avignon was full of our performances from the Festival of India that year.

The *India Today* excerpt stated: "Buoyed by the spectacular success of the Mela, both delegations seemed to be in an upbeat mood when they met to discuss the more serious affairs of state. The French made an exceptional effort to treat Rajiv as a 'very special guest'. Contrary to normal practice, Prime Minister Fabius was present in person to receive him at the airport — a job usually entrusted to a Minister. Rajiv and Sonia were lodged at the opulent Palais Marigny, a stone's throw away from the Elysee Presidential Palace and now used to house sovereigns and selected heads of state.

"Rajiv Gandhi met Mitterrand four times and yet again, as a special gesture, Mitterrand attended a dinner at Indian Ambassador Air Chief Marshal I.H. Latif's residence hosted by the Prime Minister".

It was an unusual honour by the President and Madame Mitterrand that they came for the dinner to our Embassy residence which was hosted by Prime Minister Rajiv and Sonia Gandhi. We were told that this was a special exception as the President apparently did not go to dinners at the residences of Ambassadors of any country. Another special honour extended to Prime Minister and Mrs. Sonia Gandhi by the French President

was that he met Prime Minister Rajiv Gandhi no less than four times during Rajiv's short visit. These meetings included an official lunch for Rajiv and Sonia, and a private breakfast, even though it was over a weekend, which the French keep strictly for the family, never undertaking official engagements over a weekend.

The article in *India Today* also stated,

> *The pomp and glitter of the elaborate French protocol was reinforced by unprecedented security arrangements. A dozen outriders were followed by carloads of special agents. Crack sharpshooters manned rooftops whilst elite anti-terrorist squads were on constant alert. The ubiquitous gendarmes were mobilised in their hundreds and the venue of every function Rajiv Gandhi was scheduled to attend was minutely searched. Even at spots where the Prime Minister was seen only for seconds while descending from his car, the secret service had the neighbourhood finely combed and watched windows through binoculars. The Prime Minister and Mrs. Sonia Gandhi appeared to have charmed both the media and the public. French newspapers eulogised the "handsome and dignified couple" and their well-behaved children.*

One commentator described Rajiv as "discreet", "direct", "practical" and one who refrained from rhetoric and moralising. He described Rajiv Gandhi as the "computer boy who had the grace of a young rajah who had abandoned his turban and his jewels". Rajiv conducted his meetings and his public engagements with commendable self-assurance. One example was at his meeting with senior editors of the Western press. Asked by a British journalist if appalling wages made Indian bureaucrats and administrators susceptible to bribes or indulging in espionage, Rajiv retorted that there was no evidence that better salaries enhanced a person's personal integrity. After all, Britain, which was known to pay its administrators well, had produced some of the greatest spies of modern history. Even the redoubtable foreign policy specialist, Marie-France Garaud, seemed visibly impressed by Rajiv.

In his speeches, Rajiv also referred consistently to Pakistan's nuclear programme. Answering a reporter of the left-wing political weekly, *Le Nouvel Observateur* as to whether relations between France and India would be affected if France helped Pakistan acquire a nuclear bomb, Rajiv responded that it certainly would since this bomb could only be used against India. Besides favourite issues

like the North-South dialogue, disarmament, and the possibility of restructuring global trade through a new international monetary conference, of immediate interest was the review of bilateral relations.

The French media was also interested in the number of helicopters and Airbuses, and Mirages 2000 that India was considering acquiring, plus the Dauphin helicopters for the Oil and Natural Gas Corporation (ONGC). There were also other queries about the 155-mm gun, which was a hot contender for our Army and, by now, in the final stages of consideration. Surprisingly (but happily), the use of French knowhow and possible participation of the French in the cleaning of the Ganga, which Idris briefly mentioned to Prime Minister Fabius on his first formal call on him, a river of enormous significance for Indians, and, therefore, a possible area for, if the French wished, assistance. At a press conference, clarification was sought by some on the Indian interest in the creation of a Centre for Advanced Research for this and related matters. Of very special interest was our Prime Minister's positive response to a question regarding French partnership in the development of the LCA, the Light Combat Aircraft, which, again, had been briefly mentioned by Idris in his initial call on their Defence Minister, M. Hernu, as one of the possible areas for future Indo-French collaboration.

By this time (i.e., the inauguration of the Festival of India), Idris had, fortunately, in spite of his complete lack of knowledge of the French language, developed good relations with those who shaped policy, both amongst the Ministers as well as the key officials of the French government. Idris often told me — and many others — that this was made possible only because of the extraordinary and transparently genuine interest and, indeed, admiration and respect, the French had for our history and our civilisation. This greatly helped Idris, particularly with the French government's transparently sincere desire for closer relations with our country and our people.

Idris was able to arrange several projects of value for the Government of India, and he had the French all set to cooperate fully. A notable example was the enthusiasm of the French Defence Minister M. Charles Hernu himself for the participation of France as a full partner in the development of the Light Combat Aircraft. What greatly helped Idris, he said to me later, was M. Hernu's warmth and simplicity. Even though Idris couldn't speak any French at all, M. Hernu, who couldn't speak English, invited Idris to have breakfast with him in his office: "Let's have breakfast together, tomorrow, if

you are free, and we can discuss it all at leisure and in peace," as the Minister put it.

Of course, Idris was delighted and promptly agreed. During breakfast, Idris explained to him at length, through an interpreter, his concept of the Light Combat Aircraft, which consequently would, Idris stressed, also be a Low Cost Aircraft. This was of key importance, he stressed, as India and many other countries, in spite of severe economic constraints, could then still afford to acquire it in substantial numbers and, with it, where needed, make its indigenous production a viable proposition. M. Hernu was clearly delighted with the concept and suggested that our Air Chief and his French counterpart should promptly be asked to evolve a Joint Air Staff Requirement. If successful, Monsieur Hernu said he would guarantee his country's full participation not only in the design of the airframe but the development of a suitable engine and the avionics. Encouraged by M. Hernu's enthusiastic reaction, Idris told me that he added one more element. He said that the final proof of true collaboration would be for both countries to agree, in principle, to place a substantial advance order, enough to equip a meaningful number of their frontline operational squadrons, and, finally, perhaps, even jointly promote international marketing.

All this, of course, was, Idris told me, understood by each of them, as a completely informal understanding; but each agreed to sound his own Air Chief and government. For Idris, a wonderful opportunity presented itself when Air Chief Marshal Katre visited France soon thereafter. Idris and Laxman Katre had been intimate friends for many years; Idris also thought very highly of Laxman as a person and as a professional. Happily, Laxman, Idris found, was no less enthusiastic and immediately agreed to speak on his return to the Defence Minister. Tragically, Laxman Katre's untimely death soon after his return deprived the project of his dynamic support. Also, Idris, to his dismay, found that other ideas and interests had clearly intervened and France was, at least seemingly, sidelined. The Light Combat Aircraft, after more than a quarter of a century, is yet to be in operational service.

An important challenge for the Embassy in general, and Idris in particular, was the period of the imminent arrival of a very senior Pakistani delegation. This was led by their Foreign and Defence Ministers, reportedly to finalise France's final response to Pakistan's request for major military assistance, including a very substantial number of high-performance fighter aircraft. To scuttle it, Idris called a meeting of his senior officers, and allocated to each the task of contacting the French

officials involved at an appropriately high level, the main thrust being impressing upon them the disastrous consequences to their relations with India of any major sales of military hardware to Pakistan; stressing, in particular, the impact it would, without any doubt, have on their trade potential with India across the board, which clearly by far, exceeded what Pakistan had to offer. Idris added that he himself would be speaking to the French Prime Minister, their Foreign Minister, their Finance Minister and their Defence Minister, with each of whom he had a very satisfactory working relationship. In conclusion, Idris and one of his officers mentioned to me much later — well after the event — that he once again strongly emphasised that when speaking with the French, it should be forcefully emphasised that India was a vast country of huge potential for French commercial sales and enterprise, a stable government and a country that had never, ever defaulted on timely payments of dues. If the French were serious about military sales to Pakistan, it would, without doubt, be the end of any meaningful commercial dealings with our country for a long time. They were free to make their choice! The strategy worked and, in fact, succeeded beyond their expectations.

Shri Venkatraman, Vice President of India being welcomed by the Mayor Mr. Chirac

Our Honourable Vice President, who was later to become the President of India, came for the concluding ceremony of the Festival of India. This was graciously hosted by Mr. Mitterrand, the President of France, at the Elysee Palace.

For Idris himself, though, his not having been given the time to learn French of which he hardly knew a word, continued to make the challenge daunting, as he often had to communicate through an official interpreter. We, however, found that wherever one went, if one made an effort to speak in French, the local people would be most helpful. Fortunately, I found myself remembering more and more of the fluent French which I used to speak with my mother when I was a child although I rarely spoke it thereafter, especially after my French mother passed away. Amazingly, I began to speak it again in France, but I was far from fluent after all these years. However, it improved and helped us when we were together socially or when we travelled, and, of course, I could manage on my own wherever I went.

We were very touched to discover that there was, truly, a deep and genuine interest in India and admiration of the French for the culture and philosophy of our country. Idris found that this proved a real source of French support and participation in the promotion and organisation and management of the countless Festival of India events across the length and breadth of the country. We even found that books written by Dr. Radhakrishnan and Pandit Nehru, as well as many of those on spirituality, had been translated into French and were widely known and popular.

On the very sad occasion of Jean Riboud's funeral after he passed away, Mr. Mitterrand, the President, also came for the funeral of his dear and old friend. We also accompanied the family to their chateau in the countryside where the funeral was held. We ourselves also arranged a special memorial service for Jean in the large hall at our residence, which was attended by many who had admired him greatly. It was a time of genuine sorrow for his great loss, for even in this short period, we had grown close in our hearts to him and to Krishna. The function, as we soon found, was very deeply appreciated by not only his family and friends, but also by the public at large.

Within days of our arrival in Paris in April 1985, both Idris and I both became enchanted with the unique official residence that our government had, we learnt, bought as the Indian Ambassador's residence. Apart from the magnificence of the

building, its location was remarkable: adjacent to the beautiful Champ de Mars park, and within a stone's throw of the Eiffel Tower.

However, within days of our arrival, we found that a great deal of work would be necessary, some urgently for not only its maintenance but also the restoration of the otherwise superb interiors of at least two or three of the important halls. What needed urgent attention was that the water heating in the mansion was not working, and within a few days, we were to receive several official guests in succession. It sounds ridiculous but the heating had not been set right before they started arriving, so we bought an ordinary electric water heating rod and buckets, and I kept heating the water in buckets for them. Then came a group of parliamentarians from India who were stopping in Paris for a few hours en route to Canada. They were coming for baths and breakfast to us early in the morning. I was up early in the morning heating water in two bathrooms with two immersion heaters, and sending the buckets of hot water to the guest bathrooms. Fortunately, our staff were excellent and helped willingly to carry the buckets. As we kept having to receive guests, it was fortunate that the water heating system was restored within a few weeks.

This beautiful mansion had been built, we learnt, by Consuelo Vanderbilt Balsan of the famous Vanderbilt family and her French husband Lieutenant Colonel Louis Jacques Balsan, who formerly served in the French Army during World War I and was a well-known philanthropist. Consuelo was earlier married to the Duke of Marlborough, a cousin of Winston Churchill. Their marriage was later annulled and she then married Colonel Balsan. They built and lived in this four-storied mansion for some years. She incorporated unusual marble with natural purple streaks in it, on many of the fireplaces. I was told that this was similar to the marble used at Marlborough Castle, which she may have admired when she lived there as the Duchess of Marlborough. It also had the most exquisite boisserie, wooden panelling with gold work and the special patina on them. These, we learnt, had been acquired and mounted here from an ancient chateau of the famous Cardinal Richelieu, originally built even earlier in the 1400s, which had been dismantled because it was getting destroyed by age. The boisserie, with its gold-worked panelling, had been bought and used in this mansion when it was built by Consuelo and her husband. The chandeliers and the wooden flooring were done in the same style as at the palace of Versailles.

beautiful bride was one of the loveliest that I have attended. It took place in a church in the countryside outside Paris, and the reception and dinner were arranged nearby at the chateau of the bride's grandparents. The reception was in their large formal hall and beyond that, one went into the garden for the formal sit-down dinner. There was a very large awning like our 'Shamianas' set up in the garden, but in deep blue with white stripes. There was a long line of beautifully laid out tables with candelabras and fine crystal for the dinner after the reception. At the church, I recall the lovely bride, who was so beautiful in her white wedding gown and with flowers all around her hair.

Another delightful evening was with le Vicomte Vogue and his charming wife, the owners of the famous Vaux le Vicomte which is visited by many tourists with its famous history of the origin of the Man in the Iron Mask. We had often taken visitors there, but on this occasion, there were just the four of us. It was a most delightful evening with them at their country home there, as they also related several fascinating incidents of historical interest connected with the history of the Vaux le Vicomte Chateau.

Mariam, Himmat and Farha at Mont Saint-Michel, France, 1987

A unique experience was when our daughter Mariam and her husband, the charming and fine Himmat Sandhu ('Sandy' to us) came with their little baby Farha. We drove to the beautiful Mont Saint-Michel. It is a city on a hill, like an island in the sea but is connected to the land by a short drive on a shallow side of the sea. Office buildings and houses are located on the lower area of the hill. A spiralling walk lined with charming shops took one up to the cathedral. The view of the sea and the sky all around added to the serenity and peace. On the way up, we stopped at a charming restaurant where we watched the chef prepare fluffy omelettes for each of us over an open fire. He beat the eggs to fluff them with an almost musical beat, quite fascinating to watch, and delicious to eat thereafter.

We later visited Mont Saint-Michel again when Ravi Shankar came there to give a special performance for the Festival of India. The cathedral on top of the hill with its vast hall was absolutely packed, and there were numerous young people who even climbed onto the window sills and sat enraptured throughout the performance of this great musician. Ustad Ravi Shankar had just recovered from a heart operation, but nevertheless he climbed the incredible number of steps leading up the hill to the cathedral on the summit of Mont Saint-Michel.

Another most memorable evening which was in Paris during the Festival of India was when the great Bismillah Khan performed as only he could, on the *shehnai*. Again, it had been arranged in a great church in Paris. Kishori Amonkar's was another most fabulous performance. The French audience would listen in rapt silence right through and then burst into applause that went on and on. For us, another beautiful evening which was arranged by the French was at the opera during the visit of our Prime Minister. We were so fortunate that we could drive all over and visit almost every region of this beautiful country and the countries nearby.

An unusual event for us was when Idris was invited as Guest of Honour to Dieppe in September 1986 for their two-day Rencontre Internationale de Cerf-Volants or International Kite Flying. This was organised by their Centre for Cultural Action along with the involvement of the Mayor and various international participants as well as from several regions of France. There were some fabulous kites from Thailand and China, and extremely long kites from Germany. India's expertise in kite flying combat was highlighted, and Idris also flew a kite to demonstrate this. It was quite a change from the usual diplomatic gatherings.

We entertained a great deal and it got really hectic at times. I remember when we had whole weeks of almost continuous functions at our Embassy Residence. There were important official delegations for which we hosted large receptions, and sometimes with a formal sit-down lunch party the next day, and then during the same week, we often hosted a formal dinner party on another day for a different group, and a breakfast as well as a cocktail party, all within one week. We cooked the food ourselves, so our chef Philip, and I too had lots to prepare. Philip, whom I had taught to cook, had been with us for years, ever since he was a young man. He had become really excellent after being with us at the Raj Bhavan in Mumbai where he learnt much more, as each chef there was outstanding in at least one kind of cuisine. Very often, our guests would ask us to call him at the end our sit-down dinner and they would all clap when he came in.

His Holiness the Dalai Lama with Idris and Bilkees at the Ambassador's residence

It was a great joy when His Holiness the Dalai Lama graced our residence at a reception we arranged in his honour. His wonderful smile spread his radiance over all of us.

Towards the end of our stay in France, we received news from Mumbai along with newspaper cuttings stating that Idris was being mentioned as the next President. He was actually being considered for the august post of President of India. There were several newspaper reports sent to us from India referring to the pros and cons about having a former military man as the President. The very fact that Idris' name was being seriously discussed was a great honour.

We never enquired further and forgot all about this. However, some time after we returned to India at the end of Idris' tenure as Ambassador Plenipotentiary to France, we were given some interesting and intriguing information about this. We were told that Idris had actually been seriously considered, but then there were other very active aspirants trying for the honour. We were told this very seriously when we

happened to meet the Minister Rajesh Pilot who was a former Air Force Squadron Leader, at an airport.

However, as I have mentioned in Chapter 1 of this book, while Idris was serving in France he was asked, in very great confidence, by the Prime Minister, Rajiv Gandhi, if he would agree to be the Vice President of India. Idris said that he was deeply, very deeply touched and highly honoured to be considered for such an office, but he sincerely felt temperamentally unsuited for the position, particularly as his insistence on a very high standard of discipline, which has become so much a part of him, may well, politically prove to be, in the Rajya Sabha, a cause of serious embarrassment for the government. I still get a wonderful feeling at the thought that he was so greatly honoured.

Lunch with Madame Mitterand at the Elysee Palace, Paris, 1986

Our last meal in Paris was with President Mitterrand's brother Robert and his beautiful wife Arlette. They had also known my father when he was the Indian

Ambassador to France from 1960 to 1963, and always referred to my father Nawab Ali Yavar Jung as "Un grand Bon-homme," or a great gentleman. Idris and I had been so fortunate to spend three years in this beautiful land, of people so deeply imbued in culture, a people with great aesthetic sense and refined good taste. We had lived in a most beautiful home too, very French but the way in which we added the Indian touches were much appreciated by French visitors too.

It was wonderful to see numerous incredibly beautiful regions of this fabulous country with its rich history, incredible natural beauty, and the fabulous chateaus and gardens which were each more special than the other. In addition, there are literally hundreds of fascinating museums, churches as well as libraries with so much to research and pore over. All three of our children and our son-in-law came with us when we also drove on a lovely trip to Alsace and visited Mulhouse and the hotel formerly owned there by my grandparents. For over a week, the children and the two of us stayed at a guest house that we rented from a Monsieur and Madame Zisser. They were a most charming couple who took us to their vineyard and winery, where we saw the process of making some of the famous wines of Alsace. We also joined in a delightful lunch party that they invited us to, where Madame Zisser made some of the special dishes of Alsace too. Having the children with us made this a very special holiday in the beautiful region of my mother's birth and early years.

It was a very special honour for me when I was also invited by the Institut de Science de l'Homme to join them as a Director for a one-month assignment. This is one of France's most prestigious institutions. While we were in Paris, I often went there to do my historical research for a book that I was writing. This wonderful invitation for me was a great honour but, sadly, after returning to Hyderabad, I was not able to return to Paris.

The inherent good taste and artistic eye of the French, their cultural upbringing are evident everywhere. We are so fortunate that Idris has photographed and made video films of so much that we saw, that we can almost relive our visits to these places when we view the films.

The following letter was received by Idris just when we left France, from the Institute France Asie:

INSTITUT FRANCE ASIE

Relations Humaines, Culturelles,

Economiques, Diplomatiques

Paris IV Sorbonne 1 rue Victor Cousin 75230 Paris Cedex 0

tel: 43 29 12 13 tel: 60 08 86 91

La Presidente:

Melahat Menemencioglu *April 20th, 1988*

To the most Honorable I.H. Latif,

Ambassador of the Republic of India

Your Excellency,

At the occasion of the "Year of India", the executive committee of the Institut France-Asie has unanimously decided, on my proposal, to elect you as a fourth member of its Honorary Committee, of which other three members are Mr. Alain Poher, President of the French Senate, the Minister of Foreign Affairs and the Rector Chancellor of the Universities of Paris.

We are happy to give you this personal mark of our profound consideration, after the ones we tributed to your country during the three years of your embassy, in organising first a performance in the Main Amphitheater of the Sorbonne with Indian artists, then a colloquium on Vishnuism in the palace of Senate, and finally an evening to the profit of Indian Red Cross, again in the Sorbonne.

Please accept our wishes, and believe, your Excellency, that I remain yours sincerely.

MELAHAT MENEMENCIOGLU

Madame Melahat MENEMENCIOGLU,

President,

Institut France-Asie,

Paris IV Sorbonne,

1 rue Victor Cousin,

75230 PARIS Cedex 05.

CHAPTER

30 A Potpourri of Continuing Concerns

Idris, 1988, back in Hyderabad

We flew back from France to Mumbai on 17th April 1988, and stayed in Mumbai until our car and baggage arrived. Asad flew down to Mumbai from Dubai to share the driving of our Volkswagon Jetta with Idris, back home to Hyderabad to Idris' mother's home. We stayed with her for a while and then moved to our own house for the first time. We settled in and lived there happily for more years than we had lived anywhere else together. Idris planted and nurtured nearly 50 trees, including some carefully selected fruit trees which bore delicious fruit and many equally carefully selected shrubs to make the rather rocky empty patch in front and behind our house into our lovely garden. I had been the family gardener through the years of Idris' working life, but now he took over, and loved it.

Idris' Strong Reservations on the Proposed Introduction of a Chief of Defence Staff (CDS)

In more recent years, a matter that very deeply disturbed Idris was the proposed

has incorporated some useful ideas that may bear examination. From my discussions in excellent schools abroad also, I came away with the feeling that specifically for character development, students are deliberately pressed into the rigours of research. A carefully structured system has also been evolved that gets students to observe, to analyse, to apply and to conclude largely on their own. Their Extended Essay is one such programme. Apart from an Extended Essay, I, personally, would include public speaking or elocution (which they, at present, do not have). It is a wonderful way, I have always felt, of developing both self-confidence and clarity of thought. Their CAS (which, as you know, stands for Creativity, Action and Service) is another worthwhile programme the merits of which you, as specialists, could perhaps consider. By globalisation, I certainly do not mean that we should introduce what international systems have found profitable. My suggestion is only that there may be useful international initiatives and experiences that we could examine with advantage, incorporating only those elements which reflect our needs, which we are convinced would be steps forward in our ethos; and practical in our own environment, and, of course, within our limited resources.

The second area (Character and Personality Development being the first), is our system of the 'one shot' assessment by the Central Board of Secondary Education, with no weightage at all to the performance of the student during the period preceding the Central Exam. This is patently unfair — indeed, it is also, I'm sure you will agree, clearly an inaccurate assessment of a student's real merit. A more comprehensive and, thus, a more accurate system of assessment needs to be thought out. Many, including students with whom I discussed this aspect in recent months, spoke with great feeling on this, some even expressed the hope that I would stress this when I spoke to you.

Finally, there is one other very important area that I would like to suggest which deserves urgent consideration, an area of very special importance for our country — something that was brought home to me, time and again, during my tours of the districts when I was Governor and which I touched upon in my opening remarks. Our present system is, we all know, almost

CHAPTER

30 A Potpourri of Continuing Concerns

We flew back from France to Mumbai on 17th April 1988, and stayed in Mumbai until our car and baggage arrived. Asad flew down to Mumbai from Dubai to share the driving of our Volkswagon Jetta with Idris, back home to Hyderabad to Idris' mother's home. We stayed with her for a while and then moved to our own house for the first time. We settled in and lived there happily for more years than we had lived anywhere else together. Idris planted and nurtured nearly 50 trees, including some carefully selected fruit trees which bore delicious fruit and many equally carefully selected shrubs to make the rather rocky empty patch in front and behind our house into our lovely garden. I had been the family gardener through the years of Idris' working life, but now he took over, and loved it.

Idris, 1988, back in Hyderabad

Idris' Strong Reservations on the Proposed Introduction of a Chief of Defence Staff (CDS)

In more recent years, a matter that very deeply disturbed Idris was the proposed

creation of the post of a Chief of Defence Staff (CDS). The proposal would replace the current Chiefs of Staff Committee (COSC) system, an organisation that, he often told me had, ever since Independence, served the country so very well in both peace and war.

As I have said, Idris had (and continues to have) very strong reservations on the CDS concept. However he, on principle, after retirement, always ascertained the views and policy of Air Headquarters at the level of the serving Air Chief before discussing any such policy matter concerning the Air Force with the PM or the concerned Minister or senior government official. In this case, however, in view of the extreme national significance of the proposal, he decided to also discuss the proposal and his own reservations in advance and in depth, primarily to ensure that his arguments against the proposed change had merit, and were not flawed. For this, he chose two of his close friends, both of whom, he knew, would be completely honest and forthright in giving their own personal assessment. One was (the late) Dr. P.C. Alexander, perhaps one of the most experienced and respected Indian civil servants and later Governor of Tamil Nadu and Maharashtra. The other was (the late) Shri Singhvi, the highly reputed legal luminary who had later also served for an extended period as our High Commissioner in the UK. To his great satisfaction, he found each was in total agreement with him. With the Air Chief and these two distinguished intellectuals fully in agreement with him, Idris felt greatly reassured.

One evening, watching the news on television, Idris was greatly disturbed when it was reported that the CDS concept had finally been approved by the government and was likely to be implemented within a month or two. The very next morning, he sat down and wrote a letter to Prime Minister Vajpayee, whom we had got to know fairly well when Idris was the Chief and Mr. Vajpayee the Foreign Minister. He then spoke to the Air Chief expressing his deep concern which, Idris knew was fully shared by him, and asked him if, in view of the urgency, he would be kind enough to arrange for his letter to be hand delivered at the PM's office, which the Chief readily and happily agreed to get done. Idris also wrote on very similar lines to Shrimati Sonia Gandhi since she was the Leader of the Opposition. A copy of the latter is reproduced below:

Views of our earlier home, Riyaaz, and the gardens in Hyderabad

Air Chief Marshal I.H. Latif

'Riyaaz'

House No.6-3-248/4,

Road No.1, Banjara Hills,

Hyderabad — 500 034.

29th May 2001

Dear Sonia,

I am sure you are well aware of the proposed appointment of a Chief of Defence Staff and the Air Force's strong opposition — an opposition that I must tell you, I fully share. It is comforting to know that a final decision has been held in abeyance and Govt. intends to consult the Opposition.

Mahatma Gandhi once said — I am told — that if you feel very, very strongly about something, it is a sin to be silent! It is, in fact, in that spirit that I decided to write to you, particularly as the Leader of the Opposition, and highlight certain aspects which I felt you may wish to consider:

Across the world, every country has a separate Army, Navy and Air Force; it is recognition of the fact that each Service represents a lifetime of specialisation. This, I so well remember, is something that Rajiv too recognised and often emphasised. The rapid advance of science and technology has only increased its importance. I submit that no matter how brilliant or experienced the CDS, he would never be able to correctly assess and advise the PM, the Raksha Mantri or the National Security Advisor/Council on his own, on the full and absolutely up-to-date capability of the other two Services and the options available, particularly in an emergency, when time is of the essence. He would have to consult the other Chiefs. If so, valuable time would be lost, especially when various options have to be quickly debated and decisions taken. In any case, even after consultation, his advice, at best, would be second-hand. To illustrate the point I have in mind:

The Army in J&K has been ambushed, and Govt. needs to quickly decide the response. The CDS, however, is a naval officer. Would he have adequate up-to-date knowledge of the capabilities and limitations of the Army formations in that area (weapons-wise, personnel-wise and tactics/strategy-wise) to give prompt and correct advice? Or the immediate potential/limitations of the Air Force, to correctly assess and advise the PM and the RM on the feasibility of various support options?

A report is received that an island of the Andamans group has been occupied and two unidentified ships have been sighted in the area. Would a CDS from the Army or the Air Force be able to quickly give the PM/RM satisfactory advice, advice that must again necessarily be based on comprehensive and up to the minute information of the capability of the sister Services?

Air violations have occurred in a sensitive area. Would a CDS from the Army or the Navy have adequate and up-to-date knowledge of the deployment, the vulnerability, the capabilities and limitations of the Air Force in that area to quickly assess and advise the PM and the RM on the full implications of the air violation, and response options available?

In every case, I submit, the specialist Service(s) would have to be consulted by the CDS. Would not direct/first-hand assessment and advice to the PM and the RM at the top level of each Service be infinitely preferable, especially when national security is at stake and time of the essence? If this be so, what is the shortcoming of the present system — a proven system, a system that has proved successful in every conflict we have faced?

There are other important reasons why the creation of a CDS would weaken, rather than strengthen our defence capability. But for the sake of brevity I felt that I would confine myself to just this one aspect, but an aspect which I deeply feel is of the most vital importance.

To:

Mrs. Sonia Gandhi,

10, Janpath ,

New Delhi

Earlier Idris had also called on Mr. George Fernandes, who was then Defence Minister, and Idris said later, that he met him cordially and gave him a welcome opportunity to express at length his views, in particular his strong reservations to the proposed CDS. Before he left, Idris was greatly relieved and comforted with the Defence Minister's assurance: "Don't worry Air Chief Marshal Latif, government will not go ahead with it unless all the three Defence Services are on board".

While Idris was Chief of the Air Staff, he often suggested to Air Marshal Lal that he should write about his experiences, very particularly, under his outstanding leadership about the Air Force's sterling performance during the 1971 conflict. Idris also offered to have an office and two officers arranged to help him. Finally, the retired Air Chief agreed and then wrote his excellent book, *My Life in the Air Force*. The following very valid and thought-provoking excerpts from pages 157 and 324 of Air Marshal Lal's book reflect his apprehensions on the concept of the proposed creation of the post of a Chief of Defence Staff (CDS). Intrigued, I wanted to know more about the subject and found that Air Chief Marshal Lal's book *My Years with the IAF* has a wealth of information. I find that it is one of the most fascinating books that I have read. The following is an excerpt from page 157 from his book, dealing with inter-Service matters including his views on the concept of a CDS. He writes:

> *Thanks to the breadth of vision of Pandit Jawaharlal Nehru and his early Cabinet colleagues, on the advice of Lords Mountbatten and Ismay, the British pattern of higher command was adopted with some local modifications. This pattern was founded on the principle that the three Services — the*

Army, the Navy and the Air Force — had an equally important contribution to make to the defence of the country, and it was axiomatic that they should work together. As the principal military advisers to government, the heads of the three Services were expected to meet regularly in the Chiefs of Staff Committee, where they could jointly consider all matters of military importance. In a departure from British practice, each of the Chiefs was also made responsible for the operational employment of his Service. In that capacity, the three Chiefs were expected to formulate defence plans jointly and individually, taking into account the total military environment and the capabilities and limitations of each particular Service. To reemphasise the equal status of the three Services and their interdependence, the chairmanship of the COS Committee went to the individual who had been Chief the longest, regardless of his rank or the numerical size of the Service to which he belonged. Furthermore, the close relationship that was intended between the Chiefs and their Services was reflected in the Inter-Service staff of the military wing of the Cabinet Secretariat that attended to their affairs. These basic concepts of interdependence and equality are fundamental to the proper working of our system of higher command. We ignore them at our peril.

From My Life in the Air Force, p. 324

Indeed, I saw in the proposed arrangement a positive danger to frank and free discussions particularly if the CDS happened to be excessively assertive and intolerant of the ideas of others. The important thing, I maintained, and do to this day, is that the Service Chiefs should function as equals and sort out their inter-Service problems, both administrative and operational, amongst themselves. Manekshaw, Nanda and I had, in fact, agreed upon that at the very outset of our respective terms as Service Chiefs, and that is the spirit in which Nanda and I worked throughout the Bangladesh war, and till the time we retired from our Services.

Within days of the United Progressive Alliance (UPA) coming to power, Dr. Alexander specially called Idris and persuaded him to seek an urgent meeting with Mr. Pranab Mukherjee who had been appointed as the Defence Minister. He

said that what Idris had discussed with him earlier was of the utmost importance and no time should be lost in conveying his reservations to him. Happily, Mr. Mukherjee, who Idris had never met or even spoken to earlier, also received him with comforting warmth and gave him a most patient hearing and seemed even encouragingly receptive. Thereafter, Idris felt that he had done all that he could "to obey the dictates of his conscience" as he put it (!), and could now sleep in peace! I wondered for how long!

That was, of course, years ago. Recently, when newspapers and the print media indicated that the principle of a CDS had generally been accepted by both government and the Defence Services, I once raised the subject again with Idris and asked him what he felt. I was happily surprised when he responded by saying that we must remember that many, many years have passed, and in a rapidly changing world, in every field of activity, new challenges emerge, and it would be foolish if these were not factored. In any case, at Air Headquarters, apart from a fine Chief, we have some very experienced and excellent officers. If the change is acceptable to them, there must be some very good reasons for it, and I can truthfully say that I am still sleeping in peace!!

Idris' deep interest, but also, sadly, his deep concern of the progress — or lack of progress (!) — of the LCA remains. After we returned from France to retire in Hyderabad, Idris had also, I know, discussed the lack of progress on the LCA with the Defence Minister who was then Shri Sharad Pawar. He had set up a Committee with Ratan Tata, Rahul Bajaj and Idris to look into the status of the project. Based on their study, Idris said that he was more strongly convinced than ever that the overall responsibility and control of the project should be with a carefully selected senior and suitably qualified Air Force officer.

Whilst on the subject of the LCA , I feel readers would find interesting the following letter which Idris wrote to Dr. Valluri of the National Aeronautical Laboratory (NAL) — among his several other responsibilities:

Air Chief Marshal Latif

'Riyaaz'

House No.6-3-248/4,

Road No.1, Banjara Hills,

Hyderabad — 500 034,

Andhra Pradesh

Date: 19th Oct. 1998

Dear Dr. Valuri,

What a pleasure to hear from you — and on a subject so dear to my heart! After all, the Air Force gave birth to it when I was Chief!

The story is truly a sad one — indeed, tragic: a great national challenge but within a few years to be saddled with problems, in a way, I cannot help feeling of our own making.

As Chief, I was absolutely convinced and firm throughout my term (and as Chief I was in a position to be so!) that — in relative terms at least — it must be a Low Cost Aircraft (LCA), as the Air Force would need substantial numbers. And it is essential also to make the project cost-effective, particularly as we had on our inventory and in our plans, other sophisticated (and expensive) aircraft for specialised tasks (i.e., the Jaguar, the MiG-25 and the Mirage 2000). An effective but simple battlefield attack aircraft with good self-defence, we stressed, no more, no less. Also, particularly, in view of HAL's track record, anything very advanced would have been too risky. About this also, all at our meetings were very clear — and throughout my tenure — very, very firm. In fact, it was on this basis that later, when I was in France, I discussed French collaboration and French Air Force partnership with Mr. Hernu, their Defence Minister, and received his enthusiastic support. I was very greatly disturbed and disappointed when, soon thereafter, I learnt that Air HQ had moved away from the original concept of an unsophisticated and low cost aircraft primarily for battlefield support, a concept that had been unanimously accepted at the final meeting at Air HQ.

All, present — MOD, Scientific Advisor to the DM, Chaiman HAL and Air HQ, Government agreed that this relatively modest concept was, in fact, inescapable not only because of the extreme paucity of funds measured against our urgent need for substantial numbers to replace the ageing fleet of MiG-21's, but also the compelling need to keep in mind the limitations of HAL's capability, factors which had been repeatedly mentioned and accepted by all including the then Chairman, HAL himself. I clearly recall also that before I retired l had urged all concerned to resist any change in the basic concept of a light (and consequently), a low cost aircraft. It was perhaps only later, after I had retired (maybe in '82-'83) that I got to know that Air Headquarters, after reference to field formations at the squadron level, had decided on a much more sophisticated LCA. I couldn't help feeling that in the reference to the field formations the compulsions that severely restricted our options were, perhaps, not adequately stressed.

In your article(s), if necessary to set the record right, feel free to quote me. Since it is Diwali, may I wish you, and all those near and dear to you, a wonderful year ahead.

With kind regards,

Yours sincerely,

(Idris Latif)

On his return to settle in Hyderabad in 1988, Chief Minister N.T. Ramarao invited Idris to be on the Chief Minister's Planning Board, an invitation that he happily accepted, mentioning, however, that if acceptable to the CM he would involve himself in only two of the areas of special interest to him and in which he felt he could perhaps make a useful contribution, namely, afforestation and education, which NTR unhesitatingly accepted. At the very first meeting, therefore, Idris brought up the great need for afforestation and the Chief Minister enthusiastically supported him and gave instructions to the concerned officials and department to formally establish a committee with Idris as the Chairperson. Taking advantage of the oncoming

monsoons, Idris tried his best to convince the Forest Department of the efficacy of aerial seeding, as he had done earlier when he was the Governor in Maharashtra, and organised a similar attempt as in Mumbai. But again, sadly, the support was, at best, lukewarm This was in spite of Idris telling them repeatedly that this system, successfully being used in many countries around the world, including China, was not to "replace" their work but, because the need was so great and so urgent, it would "supplement" their efforts. The representatives of the Forest Department, clearly still not convinced, raised problems of financing the experiments. When asked for a rough estimate, a figure of Rs. 20 lakh was mentioned. Idris said that it would be his responsibility to get all of it from the Centre, and was able to get double that amount for this. To his great disappointment, even the success of the initial trials did not convince them.

As aerial seeding has succeeded phenomenally in several other countries, this opposition in our country seems really inexplicable. Idris has continued to motivate those in charge, even at the Centre, and received a positive and enthusiastic response several times. Sadly, however, on each occasion, it has been opposed by those responsible in the Forest Department; truly tragic for our country with its vast drought-prone regions.

50th Anniversary of Independence: Idris' Proposal

Whilst in retirement in Hyderabad, Idris was invited by the central government to be on the National Committee to help formulate plans for the celebration of the 50th anniversary of India's Independence. After much thought, Idris decided to propose that a major event could well be something that would help heal the serious damage that had been caused by the destruction of the Babri Masjid, to the secular social fabric of India, an element so well underlined in the Constitution itself and, even more important, something so widely practised by its many communities. Idris' proposal was that at the site of the destroyed Babri Masjid, a large and beautifully designed multi-religious centre be planned, a Centre with a central hall in which the most comprehensive literature on every single religion practised in our country would be made available to the public. Importantly also, the Centre could have specially designed places for worship not only for the two affected communities but for all communities resident in India, to offer their prayers and, should they so wish, when

visiting the Centre, learn about the essence of other religions. Idris told me that he took care to emphasise that since the kind of elaborate structure he had in mind could not, obviously, be completed in time, as an event for the celebrations, a well-publicised Foundation Laying Ceremony could be considered. And for this, again, a nationally recognised and respected person from every religion practised in our country could be invited to participate.

Recognising that such a proposal may not be favoured by some, Idris mentioned to me that, on arrival in Delhi for the meeting, he took the precaution of consulting two or three senior and eminent government officials that he knew well and whose advice he valued. He was delighted that all three most strongly favoured the proposal's inclusion. Sadly, however, at the meeting the following morning, two or three very senior members for the committee felt that it was too controversial an issue to merit its inclusion, and this was accepted by the Chairman. Idris was, of course, greatly saddened. But, after the meeting, the senior government officials whom he had earlier consulted, sought an early meeting with him before his departure. When they met, they almost pleaded with him to pursue the matter, if necessary by meeting the Prime Minister, Mr. Vajpayee himself. Since the latter had always been very warm and friendly with us when Idris was the Air Chief, Idris agreed. He was happy when, in response to his request, Mr. Vajpayee gave him time that very afternoon, and when they met, Idris said, the Prime Minister gave him a very patient hearing and was happy to find that he did not dismiss his proposal outright. After quietly thinking it over for some time, he remarked: "*Haan, kheyaal tho achha hai; balke, bahut achcha hai,*" but after a pause, added, "*Maggar, mamila pecheeda hai... is par vichaar karna ho ga.*" ("Yes, the idea is good, in fact, very good, but the matter is also complex and requires very careful consideration".)

Sadly for Idris, ultimately, the committee's decision stood and his proposal was not pursued. "But," Idris added with a smile when narrating it to me, "at least I tried my best!"

How the PAN Card Came About!

Idris has periodically made unusual but important suggestions to those in the government, the President and even to Secretaries to the government. Some, encouragingly, have been accepted. Not all are known to me as they were confidential

as well as of national importance. However, one that can be written about refers to the tax system. Several years ago, Idris and I were watching a very convincing presentation by Shri N.K. Singh on some new reforms being considered for income tax returns. I cannot recall his designation at the time, but I think that he was the Finance Secretary at the central government. Idris was impressed by what he said and a thought occurred to him of a possible measure that would tighten the net around those evading taxes. This was by introducing a measure by which any purchase or expense incurred by any individual, above a relatively large amount, which in the view of the government is substantial, could only be permitted/made if the individual's income tax number is also declared and noted. And this should also apply to transactions such as settlement of hotel charges, purchase of a car, etc.

The following day, Idris, after complimenting Mr. N.K. Singh on the telephone on his fine presentationon, wrote to him on these lines. Promptly, Mr N.K. Singh responded and, after thanking Idris, stated that he would very soon see movement on his proposal on the suggested lines. Within months, the PAN Card was introduced!

During our years in Hyderabad, Idris, apart from always taking a keen interest in current affairs, such as defence, particularly anything that concerned the Air Force, was invited to be on the Governing Board of several interesting institutions. Of them, he agreed to be on the board of very few, one of which was the Court of Governors of the Administrative Staff College of India, conveniently located in Hyderabad, which he found particularly stimulating and satisfying. Earlier, on one of our visits to Mumbai, whilst staying at Raj Bhavan, in response to the invitation of the then Governor of Maharashtra, Shri Subramanium, (under whom, incidentally, he had served as the Air Chief for about a year, and whom he greatly respected), he agreed to be on the Central Executive Committee of the Bharatiya Vidya Bhavan, with its headquarters in Mumbai. Though the headquarters of the Bharatiya Vidya Bhavan is still based in Mumbai, it has, as is well known, schools spread all over India and also well known cultural centres in several other countries. However, it was their educational programmes that were of very special interest to Idris as he, I know, has always felt that though the right education is of vital importance for the future well-being and prosperity of our country, one or two areas in particular, deserve urgent attention. Some years ago, he was, therefore, very happy when, in addition to his responsibilities on the Executive Committee, he was asked to address a conclave of Bharatiya Vidya

Bhavan teachers from all their schools in our country. His address, in which he spoke frankly of his reservations, is reproduced below:

Modernisation of our System of Education

I am truly happy to have an opportunity to speak to such a distinguished gathering of Principals, especially as I have some concerns about the current education system in our country, and I can think of no better group of people to share them with, than you. As Principals of the Bhavan's institutions, for me at least, you carry a special badge of excellence! I know the wonderful work — truly wonderful work — that each of you, and all of you are doing because at every single meeting of the Bhavan's council it is the Shiksha Bharati's Activities Report that I look forward to the most.

With my Air Force background, you might think this interest of mine somewhat unusual, but it is not. It will interest you to know that for many, many years, the most important welfare activity we promoted in the Air Force was the establishment of at least one school on every Air Force camp throughout the length and breadth of our country, to cater for the children of our men in uniform. For many, many years, we did this entirely on our own — from our own private funds contributed by officers and men. This was of special importance as most of our bases have always been at far-off locations and, naturally, for all of us, the education of our children was of great concern. Today, happily, we have Kendra Vidyalaya Schools, and they are excellent. Then, later, after retirement, when I was in Mumbai as Governor — and, therefore, as Chancellor, of all my responsibilities, my involvement in the field of education was amongst the most challenging, the most stimulating — and, emotionally at least, the most satisfying.

It is with this background, and also of some first-hand and fairly recent exposure to our own and some international systems, and also with a genuine interest in this field of activity that I want to share a few thoughts with you.

The power of the pen is universally acknowledged — for teachers particularly, it has, from time immemorial, been a key instrument. Half in jest, I once

mentioned to a friend that perhaps the pencil is an even greater invention because, with an eraser, it implies an acceptance of the need — indeed, often, the importance of change: erasing a thought and substituting it with a new idea. Of course, whilst change may not necessarily be progress, PROGRESS, WITHOUT EXCEPTION, IMPLIES CHANGE. It is, you will agree, an essential ingredient. But its acceptance of the need for change demands wisdom, demands strength, demands humility; and through the ages, few sections, if any, of our society have shown these qualities in greater measure than those of our teaching community.

You know best the importance of not being bound by the past; the importance of feeling free to think freely. And I know this first-hand: when I finished my schooling and was preparing for college, I gave my autograph album to my principal to write a word of advice and guidance. Without hesitation, he wrote, "Good, better, best, let us never rest; till our good is better; and our better is best!" In short, never be satisfied! — There is ALWAYS ROOM FOR IMPROVEMENT. Never feel bound by the past; always feel free to think freely! My distinguished teachers, it is in that spirit that I would like to share a few thoughts with you.

I do not think that anyone can deny that our system of education, as given shape and substance over the years, has made giant strides and proved its merit — nationally AND internationally. Even in the most challenging fields, our young boys and girls, men and women, are second to none: the Arts and Humanities, Science, Mathematics, Medicine; in every specialisation, they have brought recognition and honour to our country. So many, many of our writers, our doctors, our scientists, our mathematicians have won national and international fame. But, as my principal said more than half a century ago (!), "Let us never rest"! Never feel bound by the past; always feel free to think freely. And in this endeavour, in our endeavour to keep pace in this fast moving field, perhaps there is merit in, so to speak, 'globalising' our view, drawing lessons from our own experience, of course, but also, perhaps, critically examining international experience and experiments. In other words, perhaps the time has come to reap the benefits of 'globalisation' that have earned rich dividends in some other fields.

Of course, the government, both at the Centre and the state level, has a primary responsibility, but our Bhavan is the premier NGO in this field, nationally — indeed, internationally recognised and truly respected. And since education lies at the very heart of the Bhavan's activities, if changes are desirable, let us — let the Bhavan, with your involvement, be at the forefront of the change.

What changes are desirable? I put it to you that we can attempt to answer this only if we take time — time to sit back, and, against the broad national horizon of our culture and our traditions, define the basic objective of education and learning. Then, and only then, can we correctly identify specific areas of weakness where changes may be desirable.

By education, what do we, as a nation, want to achieve? Is it good enough that we continue to produce great or even outstanding scientists, or mathematicians, or doctors or writers? Is education only about providing a good means of livelihood? Or, at best, achieving excellence in various fields of specialisation? Nothing wrong in that; but is that all? Surely that is not all? Surely, it is more than that? And teachers like you understand this best; you understand this best because your very profession not only encourages you but also enables you, like parents, to take a broader and deeper view of human conduct and its important elements.

Searching for a possible answer, searching for a possible broad objective, I was struck by what a very distinguished citizen of our country who, happily for us, is also a very distinguished member of our Bhavan family, Dr. Singhvi said in his Kothari Memorial lecture earlier this year, and which, happily, has been reproduced in our journal. He said, and I quote: "Knowledge must be formulated into (a) philosophy of life. (And) Philosophy must be translated into conduct and character".... I repeat "a philosophy of life translated into conduct and character". Is it not true? "A philosophy of life translated into conduct and character". He goes on to emphasise that education is (certainly) meant to advance the frontiers of knowledge and philosophy (but) it is also meant to build character...... to inculcate moral commitment and to inspire humanity to a higher purpose. Mahatma Gandhi,

totally urban-oriented. At this stage of our development, when perhaps 70% of our population is still in the rural areas, our system still pays little attention to their needs, their capacity or their genius. Something definitely needs to be done: focus on their needs, stimulate their interest and develop their expertise in the world to which they belong — a world which is also of such great importance to the overall prosperity of our country. I have, in my opening remarks, already given you an idea of how unsuited and indeed unfair the present curricula and textbooks are to the needs and capabilities of the rural population. I urge you all, all of you who are the custodians of a vital field of national activity, to give thought to the imperative need of developing, possibly, a well-structured stream within our system of education that will better serve the voiceless millions in our rural areas, millions who, in a way, give us all our daily bread.

You may well agree with some of the points that I have made, but in your minds wonder why these should be addressed to you. You may well say that, after all, these are matters of national policy and, therefore, best taken up with the concerned departments of the Centre and the states. But I have a purpose: The role of our teachers — certainly of the principals, and very specially of the standing of the institutions of the Bharatiya Vidya Bhavan, your role and your efforts, perhaps, must no longer, I truly feel, be confined to achieving excellence in our schools and colleges. Do not organisations such as FICCI and CII, for instance, very effectively function as 'pressure groups' and advise governments and help shape policy?

The Bhavan is, as I have said, the premier NGO of the country in the field of education. In fact, quite recently, you will recall, we had the unique privilege of having as our President, no less a person than a former President of the country. I wonder if any other NGO has ever had that distinction in our country. As distinguished members of the Bhavan's family, as our principals specially, should your responsibilities end in achieving excellence in your respective schools and colleges? Would you not happily consider a wider responsibility? We are all part of the great Bhavan family. Education is the very heart of our activities. Surely, we have a wider role to play? A greater responsibility to shoulder? Should we not use the standing of the Bhavan

and offer to work with the government in a task of such great national importance? It is in that spirit and with that objective that I look to you, to share your experience, your time, your energy and your invaluable expertise. It is with that objective in view that I have proposed to the Executive Committee of our Bhavan to form a small Sub-Committee to examine and recommend a course of action for the modernisation of our country's system of education. Should the Executive Committee of our Bhavan favour my suggestion, I am sure I can count on your support in evolving meaningful proposals.

We continue to go to Delhi for the annual Air Force Day celebrations to which every Air Chief has caringly and thoughtfully sent us invitations. It means a great deal to both of us to be there on these occasions. We get a chance to meet old friends and colleagues, and join in all the functions such as the fine parade with its eagerly awaited flying display and the Air Chief's wonderfully organised and hosted evening gathering at Air House. In Hyderabad, too, we go for the Air Force functions whenever possible. By now, Idris reminds me, he has been to no less than 36 of them.

Sadly, Idris' mother passed away soon after we returned to Hyderabad to live here in our house Riyaaz, which we started building, incidentally, in early 1960, but, because of our transfers and travels, never had an opportunity to live in till we finally returned to India from Paris in early 1988. My brother died from cancer some years ago. Idris' two (of three) surviving brothers and two (of three) surviving sisters also passed away within a few years, leaving Idris, in a sense, the senior-most of his family in Hyderabad. For both of us, there was no option but to mentally adjust ourselves. These tragedies apart, we are well-settled and truly happy to remain in Hyderabad, particularly as we had the lovely house that we had built in late 1959 to move into. But in early 2003, due to advancing age and deteriorating civic maintenance standards, we decided to move to a lovely flat in the Shangrila Apartment building, fortunately, still on Banjara Hills, where, by the grace of God, we are happily settled and content.

Now We Have Reached the Year 2013, and Idris' 90th Birthday is Nearly Upon Us

What has been especially enjoyable for us both is our lovely home and garden and our times together, occasionally, just the two of us going off for picnics; but also

as mentioned in our latest Bhavan's journal of 30th June, as always, put it in direct and simple terms: "Education", he said, "does not mean a knowledge of letters; it means character building, it means a knowledge of duty".

If we agree, in essence at least, with the basic objective so simply but so clearly stated by Mahatma Gandhi, and so well amplified by Dr. Singhvi, let us attempt to identify the specifics of our present system which perhaps need re-thinking. First and foremost, I would like to suggest to you that our CBSC system of final — and I repeat FINAL assessment of each student, remains an assessment of pure academic performance; sidelining almost totally, character and conduct, so central to our basic objective. And, as a consequence, most of our teaching time and effort is concentrated on academic excellence; character and personality development finds little place, and certainly NONE in the final central assessment of each student. I know that many, if not most, of our schools recognise the extreme importance of activities that promote all round development of the students — their character, their personality, their outlook, their conduct; and have developed wonderfully conceived and innovative programmes that are part of their regular daily schedule of instruction or activities.

However, the point that I would like to put to you for your consideration is that until and unless these programmes, programmes for character and personality development, unless and until they become a COMPULSORY part of the curriculum, given due weightage AND find an important place in the final assessment, the step will have little or no value. But, with such elements of the curricula, an impartial and uniform system of assessment will also need to be evolved. Only as an example, if field games are made compulsory, the focus must obviously not be on winning but, far more important, the emphasis must be on the development of team spirit, on camaraderie, on leadership. But, let me repeat, unless such activities are made compulsory and given due weightage in the final assessment, the whole effort gets, not just seriously, but, I suggest, fatally degraded.

Towards development of character and personality, I found that the International Baccalaureate System of both training and examinations

has incorporated some useful ideas that may bear examination. From my discussions in excellent schools abroad also, I came away with the feeling that specifically for character development, students are deliberately pressed into the rigours of research. A carefully structured system has also been evolved that gets students to observe, to analyse, to apply and to conclude largely on their own. Their Extended Essay is one such programme. Apart from an Extended Essay, I, personally, would include public speaking or elocution (which they, at present, do not have). It is a wonderful way, I have always felt, of developing both self-confidence and clarity of thought. Their CAS (which, as you know, stands for Creativity, Action and Service) is another worthwhile programme the merits of which you, as specialists, could perhaps consider. By globalisation, I certainly do not mean that we should introduce what international systems have found profitable. My suggestion is only that there may be useful international initiatives and experiences that we could examine with advantage, incorporating only those elements which reflect our needs, which we are convinced would be steps forward in our ethos; and practical in our own environment, and, of course, within our limited resources.

The second area (Character and Personality Development being the first), is our system of the 'one shot' assessment by the Central Board of Secondary Education, with no weightage at all to the performance of the student during the period preceding the Central Exam. This is patently unfair — indeed, it is also, I'm sure you will agree, clearly an inaccurate assessment of a student's real merit. A more comprehensive and, thus, a more accurate system of assessment needs to be thought out. Many, including students with whom I discussed this aspect in recent months, spoke with great feeling on this, some even expressed the hope that I would stress this when I spoke to you.

Finally, there is one other very important area that I would like to suggest which deserves urgent consideration, an area of very special importance for our country — something that was brought home to me, time and again, during my tours of the districts when I was Governor and which I touched upon in my opening remarks. Our present system is, we all know, almost

the company of our wonderful close family members and some very good friends. Invariably, however, the primary focus of our thoughts has been our children's visits to us, our days, sometimes even weeks, together; and our visits to them in their homes abroad. They have been in different countries, Nepal, China, the US and Dubai in the UAE. Happily, after the initial few years, our son Asad, after his marriage, decided to return to India, built a beautiful home barely minutes away from us, and happily lives here with his caring wife Huma and their very loving son Rahim, now nearly twelve years old. Mariam and Asgar, though far away now — in Dubai and New York — and travelling for us, with our advancing age, becoming difficult, very caringly find time and make the effort to visit us several times for a few days every year. Truly, God's great gift to us.

Another particularly enjoyable annual visit for many years till recently was to the seaside at Kihim, a coastal village in Alibagh District of Maharashtra. Idris' father and other relatives had bought plots along the coast of this village, and have been going there for holidays for over a century. They had all built houses, some large, some small, along with coconut plantations just in from their beach.

Hadia, Kihim

One such house with its coconut grove, 'Hadia', was being sold by a close cousin, and our three children pooled together and bought it some years ago. While our daughter was in Mumbai where her husband Himmat Sandhu was the General Manager of the two Oberoi Hotels, they often went there and our daughter looked after the property. Some years after we came to live in Hyderabad, however, Mariam and her family moved from Mumbai, to Kathmandu and on to Shanghai. Our two sons, however, were still based in New York so, primarily to look after their property, we made regular annual vists to Kihim, not only to to see to the upkeep of 'Hadia', its plantation and house; but also because we both just loved our visits there. The lovely beach and the cool breeze at the house as we would sit on the terrace looking out at the beautiful sunset over the sea brought an inner peace. The visits also gave us ample time to ensure proper maintenance of the property. The coconut water from the twenty-odd coconut palms was the sweetest coconut water that we have ever had. Unfortunately, in recent years, we found the journey too tiring, and were unable to go there, and the children rightly decided to sell the property, happily to a close and dear friend of us all.

After a delightful visit by Air Chief Marshal and Mrs. Satya Mehra to our home here, Omee Mehra wrote the following message in our Visitor's Book which we deeply treasure:

> *It has been my privilege to have worked with Idris for over three decades during which period we have shared joys, sorrows, trials, tribulations, successes and, above all, an enduring friendship and understanding. That is a fact of life which is deeply embedded in my mind and I feel certain is in that of Idris' sharp intellectual recesses. The phase which has assumed greater consequence and meaning was the one when we each acquired a partner who stood steadfastly by us and helped us along in building bridges of friendship.*
>
> *I would be failing in my duty if I failed to record that Satya and I have enjoyed the association, nay friendship of the Latifs always, and that too in plenty. This house has always welcomed us with open arms, displaying genuine warmth and affection, which has left the deepest impact on the two of us. It is to the credit of these two, God's own people that they have*

passed on their qualities to their flesh and blood — their children. God bless them and make them worthy of this charming couple.

I hope we have been, and if so, will continue to remain, worthy of your affection, regard and friendship. May God shower His choicest Blessings on you both and your near and dear ones is our fervent prayer so that this house remains full of cheer, happiness and a model of hospitality, love, affection and endless comings and goings on the part of your friends.

Omee

I was and am still very involved with the women, children and youth of several of Mumbai's largest slums and 30 tribal and other villages of Thane District. However, I was also requested for two five-year tenures to be the Chairperson of the National Bal Bhavan, of which Mrs. Indira Gandhi had been the first Chairperson. During the first tenure, we were able to establish a national award in four categories for highly talented children. In the second tenure, the Director and staff worked very hard and we established 100 more Bal Bhavans taking the total to 168 and 76 Kendras. I was able to introduce "Values" in the Delhi Government Schools, discussions on it in 180 other schools and an award for the Bal Bhavans in this. This meant a lot to me. My three earlier books were written through the years, of which *Forgotten* has been for a while on the bestseller list in Hyderabad and in Delhi. The French translation of my earlier one about my mother *L'Inde ou Vecut ma mere* was on the bestseller list in France.

While we were still in Mumbai and then later on also, I had received a few awards, including the Lions International Medal for the rehabilitation work after the Mumbai and Bhiwandi riots. I received the Padma Shree for Social Work in 2009 and was overwhelmed to get this National Award. Later, I also received the Chishti Award for Communal Harmony in 2011 at a lovely function in Delhi, when it was handed to me by the Honourable Vice President Shri Hamid Ansari.

Idris and I are so deeply grateful to God that we have three children of character and fine values — Mariam, Asad and Asgar. Mariam with her husband Himmat Sandhu (Sandy) and children Farha and Noor; Asad with Huma and son Rahim; and Asgar with Fozia and their children Nabeela and Laila. Each one with truly fine qualities of head and heart: loving, caring, intelligent and hard-working. Each is very gifted, in

Smt. Sonia Gandhi at a function at the National Bal Bhavan with the Chairperson Begum Bilkees Latif, New Delhi, 1997

Begum Bilkees receiving the INTERNATIONAL LIONS AWARD

Index

JS 214